ANALOG CIRCUITS AND SIGNAL PROCESSING

Series Editors:
Mohammed Ismail. The Ohio State University
Mohamad Sawan. École Polytechnique de Montréal

For further volumes:
http://www.springer.com/series/7381

Manar El-Chammas • Boris Murmann

Background Calibration of Time-Interleaved Data Converters

Manar El-Chammas
Texas Instruments, Inc.
12500 TI Blvd.
Dallas, TX 75243
USA
manar.chammas@gmail.com

Boris Murmann
Department of Electrical Engineering
Stanford University
420 Via Palou, Allen-208
Stanford, CA 94305-4070
USA
murmann@stanford.edu

ISBN 978-1-4899-9462-2   ISBN 978-1-4614-1511-4 (eBook)
DOI 10.1007/978-1-4614-1511-4
Springer New York Dordrecht Heidelberg London

Printed on acid-free paper

Springer is part of Springer Science+Business Media (www.springer.com)

To our families

Preface

High-speed analog-to-digital converters have become essential components of all communication systems. While we typically think of information as sequences of discrete digital symbols, the behavior of the transmission channels in all modern systems does not conform to this abstraction. Especially when a data link is pushed toward its limits, the received signals are a complex mix of wanted and unwanted analog waveforms that must be disentangled by ever more complex equalization (and channel selection) schemes. In wireless and long-distance wireline communications, these receive-side signal processing tasks have long been dealt with in the digital domain. Only recently, however, digital-domain equalization has also gained momentum in short-distance wired links providing up to several tens of gigabit/second connectivity between computer servers and their constituent components. When going digital, the designer of these links can reap the benefits of improved programmability and increased filter lengths. On the other hand, the burden is now placed on the analog-to-digital converter, which must now be inserted to finely digitize the incoming analog waveforms in order to make them fit for digital interpretation.

This monograph captures the state-of-the-art knowledge on how such high-performance converters can be realized in modern CMOS technology. Specifically, it describes how the core concepts of time-interleaving and mismatch calibration can be leveraged to achieve energy efficient conversion at sample rates of 10 GSample/second and beyond. In the discussed implementation of a 5-bit, 12-GSample/second analog-to-digital converter, several modern and innovative enhancement techniques are employed. The first is a novel statistics-based timing calibration technique that aligns the sampler timing in the ADC's input path to within a fraction of 1 ps (the time it takes for light to travel about 0.3 mm!). The second is a device-offset calibration scheme that leverages the integration density of nanometer CMOS by employing about 250 auxiliary D/A converters for component trimming. In combination, these techniques have yielded one of the most efficient data converters for high-speed links published to date.

Composed with a well-balanced mix of theoretical analysis and practical design guidelines, this book will be a valuable resource for any circuit designer active in the development of high-speed interfaces.

Stanford, CA   Boris Murmann
August 2011

Contents

1 Introduction 1
  1.1 Overview 1
  1.2 Chapter Organization 4

2 Time-Interleaved ADCs 5
  2.1 Modeling the Time-Interleaved ADC 5
    2.1.1 Frequency Domain Analysis 7
  2.2 The Effect of Time-Varying Errors 10
    2.2.1 Frequency Domain Analysis 10
  2.3 Quantitative Error Analysis 14
    2.3.1 Error Analysis Method 15
    2.3.2 Impact of Offset 18
    2.3.3 Impact of Gain 18
    2.3.4 Impact of Timing Skew 19
    2.3.5 Simulation Examples 23
  2.4 Summary 30

3 Mitigation of Timing Skew 31
  3.1 Bounds on Timing Skew 31
  3.2 Sources of Timing Skew 32
    3.2.1 Transistor Variations 32
    3.2.2 Trace and Load Variations 33
    3.2.3 Cumulative Effects of Variations 34
  3.3 Timing Skew Mitigation 35
  3.4 Background Timing Skew Calibration 38
    3.4.1 Calculating the Correlation 39
    3.4.2 Maximizing the Correlation 40
    3.4.3 Simplifying the Algorithm 40
    3.4.4 Calibrating All the Sub-ADCs 43
  3.5 Algorithmic Behavior 45
    3.5.1 Convergence Speed 46

        3.5.2   Conditions on Input Signal   49
        3.5.3   Effect of Quantization   50
   3.6 Summary   51

4   Architecture Optimization   53
   4.1 Power Dissipation   53
        4.1.1   Dynamic Comparator First-Order Model   54
        4.1.2   Dynamic Comparator Power   56
   4.2 First-Order Optimization Framework   57
        4.2.1   Performance Limits   58
        4.2.2   Optimization Analysis   58
   4.3 A Circuit-Oriented Optimization Approach   62
   4.4 Summary   63

5   Circuit Design   65
   5.1 The Sub-ADC   65
        5.1.1   Bootstrapped Track-and-Hold   65
        5.1.2   Comparator Design   69
        5.1.3   Resistor Ladder   73
        5.1.4   Wallace Encoder   73
   5.2 The Delay Line   74
        5.2.1   The Delay Cell   75
        5.2.2   Cascaded Delay Cells   76
   5.3 Phase Generator   77
   5.4 Output Buffers   77
        5.4.1   Level Converter   78
        5.4.2   LVDS Driver   78
   5.5 Summary   79

6   Measurement Results   81
   6.1 Test Setup   81
        6.1.1   Device Under Test   81
        6.1.2   Printed Circuit Board   81
        6.1.3   Data Capture Cards   83
        6.1.4   Computer   83
   6.2 ADC Measurement Results   84
        6.2.1   Static Performance   84
        6.2.2   Timing Skew Calibration   84
        6.2.3   Dynamic Performance   88
        6.2.4   Performance Summary   89
        6.2.5   Comparisons   90
   6.3 Summary   92

7   Conclusion   93
   7.1 Summary   93
   7.2 Future Work   94

A   Wide-Sense Cyclostationary Signals 95
   A.1   WSCS Example 96

B   Comparator Power Model 99

C   Optimizing a Transistor-Level Comparator 105

D   Comparator Skew 109

E   Calculating Residual Timing Errors 113
   E.1   Residual Timing Skew 113
   E.2   Estimated Jitter 114

References 115

Index 121

List of Figures

Fig. 1.1   (a) Backplane with transceivers and data path.
            (b) Communication system model 2

Fig. 1.2   (a) Single transmitted symbol. (b) Received symbol
            with slow data rate. (c) Received symbol with fast data rate 2

Fig. 1.3   (a) Series of transmitted symbols. (b) Received
            symbols with slow data rate. (c) Received symbols with
            fast data rate 3

Fig. 2.1   (a) Time-interleaved ADC. (b) Sampling edges of
            sub-ADC clocks.. 6

Fig. 2.2   Input signal DTFT example 9

Fig. 2.3   Plotted DTFT of (a) a sub-ADC output and
            (b) the time-interleaved ADC output................................. 9

Fig. 2.4   Gain, offset, and timing skew in an N-channel
            time-interleaved ADC... 10

Fig. 2.5   Effect of mismatch on sampled signal with $N = 2$.
            (a) With no mismatch. (b) With offset mismatch.
            (c) With gain mismatch. (d) With timing skew 11

Fig. 2.6   Time-interleaved ADC output with offset mismatch 13

Fig. 2.7   Time-interleaved ADC output with gain mismatch 13

Fig. 2.8   Time-interleaved ADC output with timing skew 14

Fig. 2.9   (a) Vector representation for sub-ADC mismatch
            assuming $N = 4$. (b) "Best Fit" vector is the *solid
            arrow*, and is obtained by minimizing the mean-square
            error with all the sub-ADC vectors................................... 15

Fig. 2.10   (a) Slow signal. (b) Wide autocorrelation for slow
            signal. (c) Fast signal. (d) Narrow autocorrelation for
            fast signal 20

Fig. 2.11 Setup for simulation examples 23
Fig. 2.12 Comparison of theoretical and simulation based
          SNR_τ with an input signal autocorrelation function of
          $R(\tau) = \text{sinc}(2f_c\tau)$, for $f_c = 0.1f_s, 0.25f_s$, and $0.5f_s$ 24
Fig. 2.13 Comparison of theoretical and simulation based
          SNR_τ with an input signal autocorrelation function of
          $R(\tau) = e^{-2\pi f_{3dB}|\tau|}$, for $f_{3dB} = 0.02f_s, 0.05f_s$, and $0.2f_s$ 25
Fig. 2.14 ADC SNR as a function of the standard deviation of
          timing skew, which is calculated using equality in
          (2.74). Input signal is bandlimited white noise and has
          an autocorrelation function of $R(\tau) = \text{sinc}(2f_c\tau)$ 26
Fig. 2.15 Comparison of standard deviation of skew for
          second-order low pass filter and sine wave, where α is
          such that $f_{3dB} = \alpha \hat{f}$ and $\beta = \sigma_\tau/\hat{\sigma}_\tau$ 27
Fig. 2.16 Autocorrelation function $R(T_0 + \tau/2, T_0 - \tau/2)$
          as a function of the sampling point T_0 and skew τ.
          Input signal is WSCS and has an autocorrelation
          function as in (A.11), with $\omega_{3dB} = 2/T$. (a) The
          actual autocorrelation function. (b) The autocorrelation
          function normalized such that $R(T_0, T_0) = 1$ 28
Fig. 2.17 Comparison of theoretical and simulation based SNR_τ.
          Input signal is WSCS and has an autocorrelation
          function as in (A.11). (a) With $\omega_{3dB} = 10/T$. (b) With
          $\omega_{3dB} = 1/T$ 29

Fig. 3.1   Bounds on the ADC resolution 32
Fig. 3.2   (a) Sub-ADC clocks created by phase generator.
          (b) Sampling edges of sub-ADC clocks 33
Fig. 3.3   (a) Two inverter chains and (b) standard deviation of
          timing skew as a function of power 34
Fig. 3.4   (a) Two inverter chains with load variations and
          (b) standard deviation of timing skew as a function
          of load variations 35
Fig. 3.5   Clock distribution circuit 35
Fig. 3.6   (a) Single sampler used for all sub-ADCs. (b) Signal
          and clock waveforms when single sampler is used 36
Fig. 3.7   Correction in the (a) digital domain and
          (b) mixed-signal domain... 37
Fig. 3.8   With foreground calibration, (a) ADC is either online
          and samples input or (b) is offline and is calibrated.
          With background calibration, (c) ADC is calibrated
          while sampling the input signal 38
Fig. 3.9   Attaching a calibration ADC to the time-interleaved array 39

Fig. 3.10 (**a**) Calculating the correlation between the calibration
           ADC and the sub-ADC. (**b**) Maximizing the correlation
           with a variable delay line 40
Fig. 3.11 (**a**) Output of single-bit calibration ADC. (**b**) Output of
           sub-ADC 41
Fig. 3.12 Correlation of single-bit outputs with offset 42
Fig. 3.13 Adding the calibration comparator to the
           time-interleaved array 43
Fig. 3.14 Timing diagrams for calibration clock and sub-ADC
           clocks. (**a**) Calibration clock with a period of $9T_s$.
           (**b**) Calibration clock with a period of $17T_s$ 44
Fig. 3.15 (**a**) Clock-gating to create the calibration clock, with
           an example control circuit for divide-by-three. (**b**)
           Using an integer-PLL to create the calibration clock 45

Fig. 4.1   (**a**) Cross-coupled inverter based dynamic latch.
           (**b**) Linearized cross-coupled inverter based dynamic latch 55
Fig. 4.2   (**a**) Optimal width for the first-order comparator model.
           (**b**) Optimal time-interleaved ADC power 60
Fig. 4.3   Smallest possible interleaving factor for a given power
           dissipation with a metastability rate of (**a**) 10^{-9} and (**b**) 10^{-6} 61
Fig. 4.4   Smallest possible interleaving factor as a function
           of metastability 61
Fig. 4.5   Optimal power with resistor ladder 62
Fig. 4.6   Simulated time-interleaved ADC power with different
           comparator sizings 63

Fig. 5.1   Prototype ADC architecture 66
Fig. 5.2   Sub-ADC block diagram 66
Fig. 5.3   Output SDR results of NMOS sampling switch with
           a 6 GHz input signal 67
Fig. 5.4   Track-and-hold schematic 67
Fig. 5.5   Track-and-hold with sampling capacitances 68
Fig. 5.6   Schematic of dynamic comparator 69
Fig. 5.7   (**a**) Dynamic comparator with offset correction. Reset
           transistors are not shown. (**b**) Segmented calibration
           DAC with relative transistor widths 71
Fig. 5.8   (**a**) Foreground offset correction. (**b**) Timing diagram
           for foreground offset correction 72
Fig. 5.9   7-3 Wallace Encoder 74
Fig. 5.10 15-4 Wallace Encoder 74
Fig. 5.11 Variable delay line consisting of cascaded delay cells 75
Fig. 5.12 Variable delay cell 75

Fig. 5.13 Delay cell with capacitive load 76
Fig. 5.14 Complete variable delay line 77
Fig. 5.15 Phase generator for sub-ADC clocks 78
Fig. 5.16 Level converter 78
Fig. 5.17 (a) LVDS transmitter. (b) LVDS common-mode
           feedback control circuit.. 79

Fig. 6.1   Test setup 82
Fig. 6.2   Die photo 83
Fig. 6.3   DNL for single sub-ADC (a) before offset calibration
           and (b) after offset calibration 85
Fig. 6.4   INL for single sub-ADC (a) before offset calibration
           and (b) after offset calibration 86
Fig. 6.5   Timing skew calibration algorithm using the gradient
           based maximizer. (a) SNDR convergence and
           (b) timing skew correction codes................................. 87
Fig. 6.6   Change in skew correction code after each calibration
           cycle for a single sub-ADC 87
Fig. 6.7   SNDR convergence using iterative maximizer 88
Fig. 6.8   Decimated output spectrum (a) without timing skew
           calibration and (b) with timing skew calibration 89
Fig. 6.9   Input frequency sweep. (a) SNDR performance with
           and without calibration. (b) SNR and SNDR curves
           with calibration.. 90
Fig. 6.10 Comparisons between ADCs with a sample rate larger
           than (a) 1 GS/s and (b) 10 GS/s 91

Fig. B.1   Currents in cross-coupled inverter based dynamic latch 100

Fig. C.1   Schematic of dynamic comparator 106
Fig. C.2   Simulated time-interleaved ADC power with different
           transistor sizings. The optimal boundary is outlined in black....... 106

Fig. D.1   Comparator clock sampling edges (a) without skew
           and (b) with skew 110
Fig. D.2   ADC ENOB as a function of the comparator skew.................. 111

List of Tables

Table 5.1   Capacitance sizing 68
Table 5.2   Full-adder operation 74

Table 6.1   Test equipment used in Fig. 6.1 82
Table 6.2   Performance summary of prototype ADC 91
Table 6.3   Published ADCs faster than 10 GS/s 92

Acronyms

ADC	Analog-to-digital converter
CML	Current-mode logic
DAC	Digital-to-analog converter
DLL	Delay-locked loop
DNL	Differential nonlinearity
DTFT	Discrete-time Fourier transform
DUT	Device under test
ENOB	Effective number of bits
FO4	Fan-out of four delay
FOM	Figure of merit
Gb/s	Giga-bits per second
GPIO	General purpose input/output
GS/s	Giga-samples per second
INL	Integral nonlinearity
LMS	Least-mean squares
LSB	Least significant bit
LVDS	Low-voltage differential signaling
Mb/s	Mega-bits per second
MSB	Most significant bit
PCB	Printed circuit board
PLL	Phase-locked loop
QFN	Quad-flat no-leads
SAS	Serial attached SCSI
SDR	Signal-to-distortion ratio
SFDR	Spurious-free dynamic range
SNDR	Signal-to-noise-and-distortion ratio
SNR	Signal-to-noise ratio
SoC	System-on-chip
WSCS	Wide-sense cyclostationary
WSS	Wide-sense stationary

Chapter 1
Introduction

1.1 Overview

In the foreseeable future, as in the past few decades, the integration of communication systems within our daily lives will continue to exponentially increase. This is partly fueled by the increasing data rates of serial links. For example, serial-attached SCSI (SAS), a computer bus used to communicate with storage devices, originally started at 40 Mb/s, eventually progressed to 3 Gb/s when it transformed into a serial link, and is now evolving from 6 Gb/s to 12 Gb/s [1]. These communication systems tend to have both the transmitter and receiver placed on a backplane, as in Fig. 1.1a, which can be represented by the simplified block diagram in Fig. 1.1b. Ideally, the received signal in communication systems is a perfect replica of the transmitted signal, such that the receiver can perfectly decode the transmitted data. However, this is not the case due to the dispersive properties of the channel, as governed by the channel frequency response. This response is a function of a number of parameters, such as the board design, trace lengths, and backplane material, and results in inter-symbol interference.

Ultimately, the effect of inter-symbol interference on the signal depends on the data rate. For example, with a channel input as in Fig. 1.2a, which has a symbol width of T_s seconds such that the data rate is $1/T_s$, the difference between the channel output and the channel input increases with the data rate, as shown in Fig. 1.2b,c. This becomes a more serious problem once several bits are transmitted. With a channel input as in Fig. 1.3a, the resulting channel output with a slow and fast data rate is shown in Fig. 1.3b,c respectively. In Fig. 1.3c, inter-symbol interference can lead to bit-errors.

In order for the receiver to make sense of the channel output and to meet the communication system's target bit-error rate, it must implement various equalization blocks [2]. The complexity of these equalization blocks increases with data rate, and results in higher power consumption. It is this increased complexity that has lead

M. El-Chammas and B. Murmann, *Background Calibration of Time-Interleaved Data Converters*, Analog Circuits and Signal Processing, DOI 10.1007/978-1-4614-1511-4_1, © Springer Science+Business Media, LLC 2012

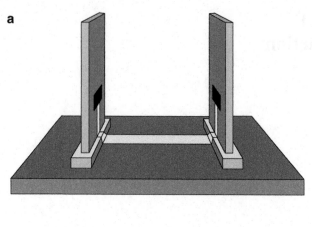

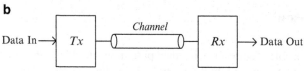

Fig. 1.1 (**a**) Backplane with transceivers and data path. (**b**) Communication system model

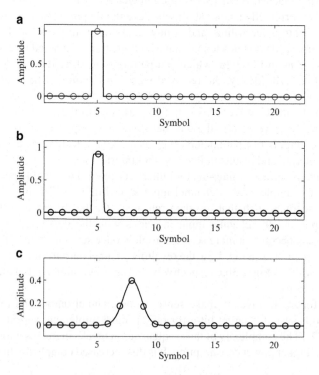

Fig. 1.2 (**a**) Single transmitted symbol. (**b**) Received symbol with slow data rate. (**c**) Received symbol with fast data rate

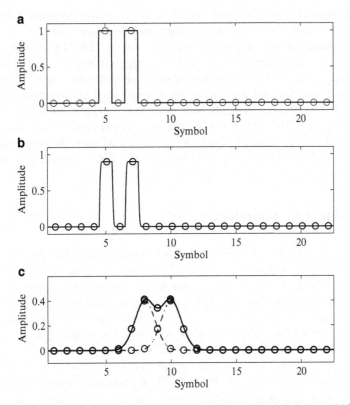

Fig. 1.3 (a) Series of transmitted symbols. (b) Received symbols with slow data rate. (c) Received symbols with fast data rate

to a trend in various wireline communication systems to use a digitally-equalized serial link, in which some of the equalization blocks are moved to the digital domain [3].

Pushing some of the equalization blocks into the digital domain increases the available design space and potentially allows for a more power-efficient partitioning of tasks in the overall system design. However, this necessitates the use of an ADC. Although the design specifications of the ADC depend on the communication system as well as on the implementation of the various equalization blocks, many wireline systems require a conversion rate of over 10 GS/s and a resolution of over 4 bits [3].

The realization of such an ADC is the motivation behind this research, as there are two main points of concern. The first is that of ADC feasibility. Currently, the fastest single channel ADC can sample at 7.5 GS/s [4], which does not meet the required specifications. The second is that of power consumption. With hundreds of transceivers located on a single board or server, a small increase in power per transceiver results in a large overhead. As a result, wireline communication systems have tight power budgets. One difficulty in creating a power-constrained digitally-equalized serial link is the poor energy efficiency of high-speed ADCs.

Both of these issues are addressed in this research. A flash ADC is used because of the low resolution requirements, and its energy efficiency is improved with the addition of hundreds of trim circuits that enable the flash ADC to meet its performance specifications while reducing its total power consumption. An ADC with a high data rate is commonly built using the technique of time-interleaving a number of sub-ADCs [5]. However, time-interleaved ADCs suffer from time-varying errors. This research proposes a statistics-based calibration algorithm that mitigates the effects of timing skew and improves the ADC's dynamic performance.

1.2   Chapter Organization

This book is divided into six chapters. Chapter 2 focuses on a theoretical overview of time-interleaved ADCs. After developing a model for time-interleaved ADCs, quantitative bounds on several time-varying errors are analyzed. Chapter 3 discusses the statistics-based calibration algorithm proposed to compensate for timing skew. Various aspects of the calibration algorithm, including convergence speed and limitations, are presented. Chapter 4 introduces a high-level optimization framework for the design of time-interleaved flash ADCs. Chapter 5 discusses the prototype ADC designed and used to evaluate the calibration algorithm, and presents the circuit techniques used to realize the ADC. The measurement results and test setup are presented in Chap. 6, while Chap. 7 draws conclusions from this work.

Chapter 2
Time-Interleaved ADCs

The time-interleaved ADC is an architecture that cycles through a set of N sub-ADCs, such that the aggregate throughput is N times the sample rate of the individual sub-ADCs [5]. Therefore, such an architecture enables the sample rate to be pushed further than that achievable with single channel ADCs. This chapter discusses the operation of time-interleaved ADCs and analyzes how the sub-ADCs interact. It also analyzes the drawbacks of the architecture and presents closed-form equations relating performance degradation to mismatch.

2.1 Modeling the Time-Interleaved ADC

This section discusses the operation of the time-interleaved ADC. The model presented serves as a foundation that allows the inclusion of time-varying errors due to differences between the sub-ADCs, as discussed in Sect. 2.2.

The time-interleaved ADC, as shown in Fig. 2.1a, has an input $x(t)$ and an output $y[n]$. The sampling period of the time-interleaved ADC and the N sub-ADCs are T_s and $\hat{T}_s = N \cdot T_s$, respectively. The ith sub-ADC, where $i = 0, ..., N-1$, is strobed with clock $\phi_i(t)$, which ideally has sampling edges at

$$t_i[n] = n\hat{T}_s + iT_s$$
$$= (nN + i) \cdot T_s \tag{2.1}$$

Thus, the sampling edges of two consecutive clocks are offset by T_s, as in Fig. 2.1b, and the input signal is uniformly sampled. The output of the ith sub-ADC is $\hat{y}_i[n]$, where

$$\hat{y}_i[n] = x\left(t_i[n]\right)$$
$$= x\left([nN + i] \cdot T_s\right) \tag{2.2}$$

M. El-Chammas and B. Murmann, *Background Calibration of Time-Interleaved Data Converters*, Analog Circuits and Signal Processing, DOI 10.1007/978-1-4614-1511-4_2, © Springer Science+Business Media, LLC 2012

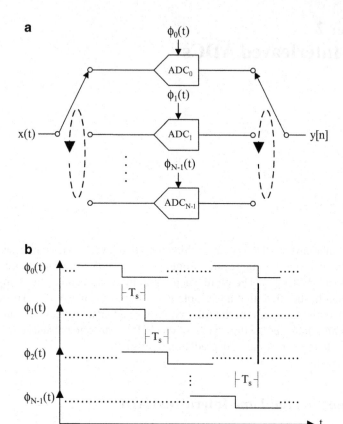

Fig. 2.1 (a) Time-interleaved ADC. (b) Sampling edges of sub-ADC clocks

The sub-ADC outputs $\hat{y}_i[n]$ are multiplexed to create $y[n]$, such that

$$y[n] = \hat{y}_i \left[\frac{n-i}{N} \right] \quad \text{where } i = n \bmod N \qquad (2.3)$$

Setting $y_i[n]$ as the sub-ADC output $\hat{y}_i[n]$ upsampled by N results in

$$y_i[n] = \begin{cases} \hat{y}_i \left[\frac{n-i}{N} \right] & \text{if } \frac{n-i}{N} \text{ is an integer} \\ 0 & \text{else} \end{cases} \qquad (2.4)$$

This is simplified by defining

$$\delta_i[n] = \sum_{k=-\infty}^{\infty} \delta[n - kN - i] \qquad (2.5)$$

such that

$$y_i[n] = x(nT_s) \cdot \delta_i[n] \tag{2.6}$$

Thus, the time-interleaved ADC output $y[n]$ in (2.3) becomes

$$y[n] = \sum_{i=0}^{N-1} y_i[n] \tag{2.7}$$

As expected, the output of the ideal time-interleaved ADC reduces to $y[n] = x(nT_s)$.

2.1.1 Frequency Domain Analysis

The discrete-time Fourier transform (DTFT) is used to represent the time-interleaved ADC discrete-time output $y[n]$ and the sub-ADC output $y_i[n]$ in the frequency domain [6]. In general, the DTFT of a discrete-time input $x[n]$ [7] is

$$X(f) = \sum_{n=-\infty}^{\infty} x[n] \cdot e^{-j(2\pi f)n} \tag{2.8}$$

where $X(f)$ is periodic with period 1. The inverse transform is

$$x[n] = \int_{-1/2}^{1/2} X(f) \cdot e^{j(2\pi f)n} df \tag{2.9}$$

2.1.1.1 Sub-ADC Output

The DTFT of the upsampled sub-ADC output $y_i[n]$ in (2.6) is

$$Y_i(f) = \sum_{n=-\infty}^{\infty} (x[n]\delta_i[n]) \cdot e^{-j(2\pi f)n} \tag{2.10}$$

where $x[n] = x(nT_s)$. By property of the DTFT [7], $Y_i(f)$ is equal to the convolution of the DTFTs of $x[n]$ and $\delta_i[n]$. The DTFT of the sampled input $x[n]$ is $X(f)$, whereas the DTFT of $\delta_i[n]$ [8] is

$$D_i(f) = \frac{1}{N} \sum_{k=-\infty}^{\infty} \delta\left(f - \frac{k}{N}\right) \cdot e^{j\left(\frac{2\pi k}{N}\right)i} \tag{2.11}$$

such that

$$Y_i(f) = X(f) * D_i(f)$$

$$= \frac{1}{N} \sum_{k=-\infty}^{\infty} e^{j\left(\frac{2\pi k}{N}\right)i} \cdot X\left(f - \frac{k}{N}\right) \tag{2.12}$$

This results in replicas at spacings of $\frac{2\pi k}{N}$ because of the subsampling behavior of the sub-ADCs. A phase-shift exists as a function of i, due to the exponential, such that, even though the magnitude of $Y_i(f)$ is the same for all the sub-ADCs, the phases are different.

2.1.1.2   Time-Interleaved ADC Output

The DTFT of the time-interleaved ADC output $y[n]$ in (2.7) is

$$Y(f) = \sum_{i=0}^{N-1} Y_i(f) \tag{2.13}$$

and, using (2.12), can be written as

$$Y(f) = \sum_{k=-\infty}^{\infty} M[k] \cdot X\left(f - \frac{k}{N}\right) \tag{2.14}$$

where $M[k]$ is defined as

$$M[k] = \frac{1}{N} \sum_{i=0}^{N-1} e^{j\left(\frac{2\pi k}{N}\right)i}$$

$$= \begin{cases} 1 & \text{if } \frac{k}{N} \text{ is an integer} \\ 0 & \text{else} \end{cases} \tag{2.15}$$

Thus,

$$Y(f) = \sum_{k=-\infty}^{\infty} X(f - k) \tag{2.16}$$

and the inverse DTFT of $Y(f)$ is $x[n]$, as expected.

2.1.1.3   Interpretation

The sub-ADC outputs in (2.12) have frequency domain replicas with spacings of $\frac{2\pi k}{N}$. Due to the phase differences between the sub-ADC outputs, which are a function of i, all replicas except those at $2\pi k$ cancel when the sub-ADC outputs

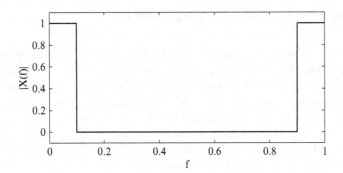

Fig. 2.2 Input signal DTFT example

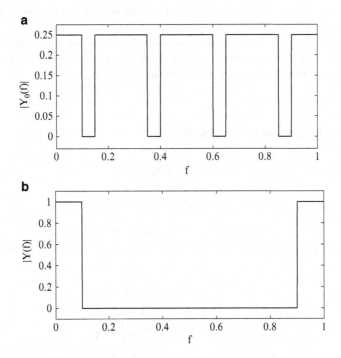

Fig. 2.3 Plotted DTFT of (**a**) a sub-ADC output and (**b**) the time-interleaved ADC output

are summed in (2.13). To illustrate this, assume that a 4-way time-interleaved ADC samples an input signal with a DTFT as in Fig. 2.2. As shown in Fig. 2.3a, the DTFT of the sub-ADC output consists of scaled replicas, whereas the resulting time-interleaved ADC output spectrum in Fig. 2.3b is identical to the input signal DTFT.

2.2   The Effect of Time-Varying Errors

As previously mentioned, and as in Fig. 2.1a, the inputs and outputs of the time-interleaved ADC are $x(t)$ and $y[n]$, respectively, where ideally $y[n] = x(nT_s)$, T_s being the sampling period of the time-interleaved ADC. Each of the N sub-ADCs is controlled by a clock with period $\hat{T}_s = N \cdot T_s$; the ideal phase offset of the clock for the ith sub-ADC with respect to the first sub-ADC is iT_s, where $i = 0, ..., N - 1$. However, as illustrated in Fig. 2.4, there are several sources of mismatch in the signal data path which degrade the ADC performance. Each sub-ADC has its own gain G_i, offset o_i, and timing skew τ_i [5], which modify (2.6) into

$$y_i[n] = \Big(G_i \cdot x(nT_s - \tau_i) + o_i \Big) \cdot \delta_i[n] \qquad (2.17)$$

for $i = 0, \ldots, N - 1$. The effect of these errors can be viewed in the time domain, as in Fig. 2.5.

This section uses the frequency domain to develop a more intuitive understanding of how the outputs of the mismatched sub-ADCs interact and the time-domain to quantify the relationship between mismatch and ADC performance.

2.2.1   Frequency Domain Analysis

The ith sub-ADC output in (2.17) can be rewritten as

$$y_i[n] = \Big(h_i(nT_s) * x(nT_s) + o_i \Big) \cdot \delta_i \qquad (2.18)$$

where o_i is the sub-ADC offset and $h_i(t)$ is a linear time-invariant function that is used to model both the sub-ADC gain and timing skew. It can also be used to

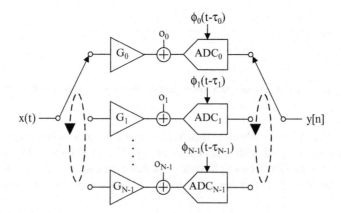

Fig. 2.4 Gain, offset, and timing skew in an N-channel time-interleaved ADC

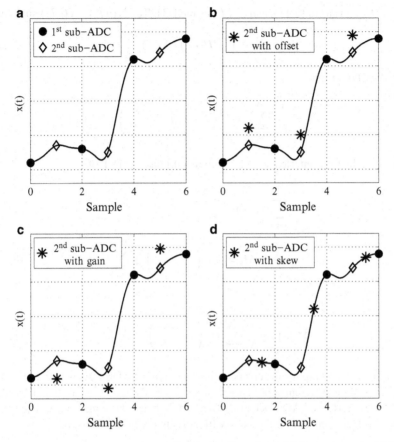

Fig. 2.5 Effect of mismatch on sampled signal with $N = 2$. (**a**) With no mismatch. (**b**) With offset mismatch. (**c**) With gain mismatch. (**d**) With timing skew

model other effects, such as bandwidth mismatch [9], although this is not discussed here. For example, gain is modeled with $h_i(t) = G_i \cdot \delta(t)$ and timing skew with $h_i(t) = \delta(t - \tau_i)$. When these effects are included, the DTFT of $y_i[n]$ in (2.12) becomes

$$Y_i(f) = \sum_{n=-\infty}^{\infty} \Big((h_i(nT_s) * x(nT_s) + o_i) \cdot \delta_i[n] \Big) \cdot e^{-j(2\pi f)n} \qquad (2.19)$$

Defining $O_i(f)$ as

$$O_i(f) = o_i \cdot D_i(f) \qquad (2.20)$$

where $D_i(f)$ is as in (2.11), and $\hat{X}_i(f)$ as the DTFT of $h_i(nT_s) * x(nT_s)$ such that

$$\hat{X}_i(f) = H_i(f) \cdot X(f) \tag{2.21}$$

simplifies $Y_i(f)$ into

$$Y_i(f) = \frac{1}{N} \sum_{k=-\infty}^{\infty} e^{j\left(\frac{2\pi k}{N}\right)i} \cdot \hat{X}_i\left(f - \frac{k}{N}\right) + O_i(f) \tag{2.22}$$

Therefore, the time-interleaved ADC output $y[n]$ has a DTFT of

$$Y(f) = \sum_{i=0}^{N-1} Y_i(f)$$

$$= \sum_{k=-\infty}^{\infty} M_h[k] \cdot X\left(f - \frac{k}{N}\right) + \sum_{i=0}^{N-1} O_i(f) \tag{2.23}$$

where

$$M_h[k] = \frac{1}{N} \sum_{i=0}^{N-1} H_i\left(f - \frac{k}{N}\right) \cdot e^{j\left(\frac{2\pi k}{N}\right)i} \tag{2.24}$$

This is a generic setup for the errors in time-interleaved ADCs. As is seen in (2.24), the phases of the different sub-ADCs do not necessarily cancel out as they did in the ideal time-interleaved ADC because of $H_i(f)$, which is no longer unity. The three cases of offset, gain and timing skew will individually be expanded on.

2.2.1.1   Effect of Offset Mismatch

With offset mismatch, $h_i(t) = \delta(t)$ such that $H_i(f) = 1$, and $o_i \neq 0$. Therefore, $M_h[k]$ in (2.24) simplifies to (2.15), and

$$Y(f) = \sum_{k=-\infty}^{\infty} X(f - k) + \sum_{i=0}^{N-1} O_i(f) \tag{2.25}$$

The resulting spectrum has tones spaced at $\frac{2\pi k}{N}$, due to $O_i(f)$. These tones are not a function of the input signal, and only depend on the size of the offsets and the number of sub-ADCs. For example, using the input spectrum of Fig. 2.2, the resulting output with an interleaving factor of four and with offset mismatch is as shown in Fig. 2.6.

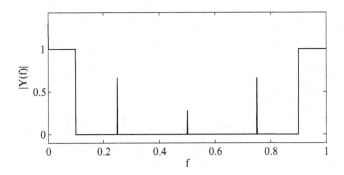

Fig. 2.6 Time-interleaved ADC output with offset mismatch

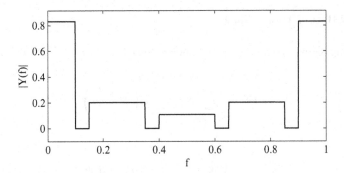

Fig. 2.7 Time-interleaved ADC output with gain mismatch

2.2.1.2   Effect of Gain Mismatch

With gain mismatch, $h_i(t) = G_i\delta(t)$ such that $H_i(f) = G_i$, and $o_i = 0$. Therefore,

$$Y(f) = \sum_{k=-\infty}^{\infty} M_h[k] \cdot X\left(f - \frac{k}{N}\right) \qquad (2.26)$$

where

$$M_h[k] = \frac{1}{N}\sum_{i=0}^{N-1} G_i \cdot e^{j\left(\frac{2\pi k}{N}\right)i} \qquad (2.27)$$

If $G_i = 1$ for all the sub-ADCs, then $M_h[k]$ becomes $M[k]$, as previously defined. However, when the gains are not all identical, the replicas in the sub-ADC outputs do not necessarily cancel out. The magnitude of these residual replicas is a function of the sub-ADC gains, such that the gain errors effectively amplitude modulate the input signal. For example, Fig. 2.7 plots the resulting output DTFT for an ADC with gain mismatch and an interleaving factor of four, using the input signal of Fig. 2.2. As is expected non-zero replicas exist because of gain errors.

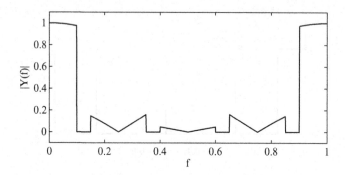

Fig. 2.8 Time-interleaved ADC output with timing skew

2.2.1.3   Effect of Timing Skew

With timing skew, $h_i(t) = \delta(t - \tau_i)$ such that $H_i(f) = e^{-j(2\pi f)\tau_i}$, and $o_i = 0$.
Therefore,

$$Y(f) = \sum_{k=-\infty}^{\infty} M_h[k] \cdot X\left(f - \frac{k}{N}\right) \tag{2.28}$$

where

$$M_h[k] = \frac{1}{N}\sum_{i=0}^{N-1} e^{-j2\pi\left(f-\frac{k}{N}\right)\tau_i} \cdot e^{j\left(\frac{2\pi k}{N}\right)i} \tag{2.29}$$

If $\tau_i = 0$ for all the sub-ADCs, then $M_h[k]$ becomes $M[k]$, as previously defined.
However, when the timing skews are not all identical, the replicas in the sub-ADC
outputs do not cancel. The phases of these replicas are a function of the timing
skews, effectively phase modulating the input signal. For example, Fig. 2.8 plots the
resulting output DTFT for an ADC with timing skew and an interleaving factor of
four, using the input signal of Fig. 2.2. In addition to having non-zero replicas, the
baseband signal is slightly distorted, which is a result of the frequency dependent
phase shifts caused by timing skew.

2.3   Quantitative Error Analysis

Analytic expressions quantifying the effect of the aforementioned time-varying
errors on ADC performance are important when analyzing the design space of
the time-interleaved ADC. This section relates the ADC *SNR* to these errors, and
provides statistical bounds on the acceptable mismatch.

Fig. 2.9 (a) Vector representation for sub-ADC mismatch assuming $N = 4$. (b) "Best Fit" vector is the *solid arrow*, and is obtained by minimizing the mean-square error with all the sub-ADC vectors

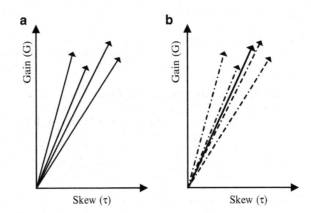

2.3.1   Error Analysis Method

Analyzing the effect of time-varying errors consists of writing the output $y[n]$ of the time-interleaved ADC in terms of two components [10] as

$$y[n] = x_o[n] + e[n] \tag{2.30}$$

where $x_o[n]$ is a uniformly sampled version of the incoming signal $x(t)$ and is the "best fit" to the time-interleaved ADC output $y[n]$ such that

$$x_o[n] = \hat{G} \cdot x(nT - \hat{\tau}) \tag{2.31}$$

and where $e[n]$ is the resulting error signal. In other words, this "best fit" is a scaled and shifted version of the original input signal. For example, if the input $x(t)$ is a sinusoidal function, then the "best fit" $x_o[n]$ in (2.31) is also a sinusoidal function and suffers from no distortion harmonics. $\hat{G}$ and $\hat{\tau}$ are derived by maximizing the output *SNR*, which is equivalent to minimizing the mean-square error, and result in $x_o[n]$ and $e[n]$ being orthogonal [11]. This method is used for all relevant mismatches, and the results obtained subsume the approach in which only a sinusoid is used as an input.

Graphically, this can be viewed with a vector space representation. Each of the N sub-ADCs is represented by a two-dimensional vector (G_i, τ_i), as in Fig. 2.9a. The "best fit" is then the vector that minimizes the mean-square error, as in Fig. 2.9b. It is interesting to note that if $G_i = G$ and $\tau_i = \tau$, regardless of what G and τ actually are, then $\hat{G} = G$ and $\hat{\tau} = \tau$, as is depicted by the vectors in Fig. 2.9b.

The input signal in this analysis is assumed to be WSS with signal power P and autocorrelation $R(\tau)$. Without loss of generality, the mean of the input signal is set to zero and mean of the sub-ADC gains is set to one, such that

$$E[x(t)] = 0 \tag{2.32}$$

and

$$\frac{1}{N} \cdot \sum_{i=0}^{N-1} G_i = 1 \tag{2.33}$$

Therefore, the mean of the error signal is

$$E\big[e[n]\big] = E\big[y[n] - x_o[n]\big]$$

$$= \frac{1}{N} \sum_{i=0}^{N-1} o_i \tag{2.34}$$

The mean-square error is defined as

$$f(\hat{G}, \hat{\tau}) = E\Big[e[n]^2\Big] - E\Big[e[n]\Big]^2 \tag{2.35}$$

such that

$$f(\hat{G}, \hat{\tau}) = \left(\hat{G}^2 P + \frac{P}{N} \sum_{i=0}^{N-1} G_i^2 - 2\frac{\hat{G}}{N} \sum_{i=0}^{N-1} G_i R(\tau_i - \hat{\tau}) + \frac{1}{N} \sum_{i=0}^{N-1} o_i^2\right) - \frac{1}{N^2} \left(\sum_{i=0}^{N-1} o_i\right)^2 \tag{2.36}$$

The mean-square error in (2.36) is minimized with respect to both $\hat{G}$ and $\hat{\tau}$ by first setting the partial derivative of (2.36)

$$\frac{\partial f(\hat{G}, \hat{\tau})}{\partial \hat{G}} = 2\hat{G} P - \frac{2}{N} \sum_{i=0}^{N-1} G_i R(\tau_i - \tau) \tag{2.37}$$

to zero. Therefore,

$$\hat{G} = \frac{1}{NP} \sum_{i=0}^{N-1} G_i R(\tau_i - \hat{\tau}) \tag{2.38}$$

This is optimal because (2.36) is convex in $\hat{G}$. The optimal "best fit" gain is thus a function of the individual sub-ADC gains and the autocorrelation of the input function.

$\hat{\tau}$ is then found by replacing (2.38) in (2.36) such that

$$f(\hat{G}, \hat{\tau}) = \frac{P}{N} \sum_{i=0}^{N-1} G_i^2 - \frac{1}{N^2 P} \left(\sum_{i=0}^{N-1} G_i R(\tau_i - \hat{\tau})\right)^2 + \frac{1}{N} \sum_{i=0}^{N-1} o_i^2 - \frac{1}{N^2} \left(\sum_{i=0}^{N-1} o_i\right)^2 \tag{2.39}$$

and (2.39) is minimized by finding the value of $\hat{\tau}$ that maximizes $\sum_i G_i R(\tau_i - \hat{\tau})$ such that

$$\hat{\tau} = \arg\max_{\tau} \sum_{i=0}^{N-1} G_i R(\tau_i - \tau) \qquad (2.40)$$

For input signals with a first-order differentiable autocorrelation function, this is equivalent to the value of $\hat{\tau}$ that satisfies

$$\sum_{i=0}^{N-1} G_i \frac{dR(\tau_i - \tau)}{d\tau}\bigg|_{\tau=\hat{\tau}} = 0 \qquad (2.41)$$

Solutions obtained with (2.41) must be checked to see if they satisfy concavity constraints for maximization.

Using the values obtained for $\hat{G}$ and $\hat{\tau}$, we can directly solve for $SNR_f = P_S/P_N$, where $P_S = P$ and where $P_N = f(\hat{G}, \hat{\tau})$. In the context of ADCs, it is meaningful to quantify the effect of mismatches by comparing the resulting SNR to that due to quantization. In an ADC with a resolution of B bits, the SNR due to quantization is

$$SNR_Q = \frac{3}{2} \cdot \left(2^{2B}\right) \qquad (2.42)$$

This is used to provide a bound on all three of the aforementioned mismatches by setting

$$SNR_f \geq SNR_Q \qquad (2.43)$$

When equality exists in (2.43), the actual SNR (which includes the effect of quantization) is $SNR = SNR_f - 3$dB. The time-interleaved ADC is "quantization-noise limited" when strict inequality exists and is "mismatch limited" when $SNR_f < SNR_Q$. This presents a deterministic bound on the relevant mismatch, and can be used to validate a given converter. In other words, given a time-interleaved ADC with a set of gain, offset, or timing skew mismatch, and given the input signal autocorrelation function it is possible to state whether the converter is quantization-noise limited or mismatch limited.

However, it is also useful to know beforehand what the acceptable mismatch is for a time-interleaved ADC with a target resolution of B bits. This can be done by bounding the variance of the time-varying error with

$$E\left[f(\hat{G}, \hat{\tau})\right] \leq \left(\frac{2}{3}\right) \cdot \left(\frac{P}{2^{2B}}\right) \qquad (2.44)$$

and by assuming that these errors are independent and identically distributed random variables. This is done for each of the time-varying errors in the sections below.

2.3.2   Impact of Offset

With the assumptions that the gain and timing skew for all N sub-ADCs are identical such that, without loss of generality, $G_i = 1$ and $\tau_i = 0$, the mean-square error in (2.39) reduces to

$$f(\hat{G}, \hat{\tau}) = \frac{1}{N} \sum_{i=0}^{N-1} o_i^2 - \frac{1}{N^2} \left(\sum_{i=0}^{N-1} o_i \right)^2 \tag{2.45}$$

Therefore, the *SNR* due to offset is

$$SNR_O = \frac{P}{\frac{1}{N} \sum_{i=0}^{N-1} o_i^2 - \frac{1}{N^2} \left(\sum_{i=0}^{N-1} o_i \right)^2} \tag{2.46}$$

and the statistical bound on the variance of offset, using (2.44), is

$$\sigma_o^2 \leq \left(\frac{N}{N-1} \right) \cdot \left(\frac{2 \cdot P}{3 \cdot 2^{2B}} \right) \tag{2.47}$$

Thus, the bound on offset is a function of the number of sub-ADCs N, the input signal power P, and the ADC resolution B. The bound on offset is unique when compared to that of both gain mismatch and timing skew since it is directly proportional to P. It is intuitive that ADCs with higher power input signals can cope with larger sub-ADC offsets. Furthermore, as shown in (2.47), higher resolution ADCs result in smaller bounds on offset mismatch, as does a higher interleaving factor, although the ADC resolution has a much larger effect on the bound. For example, if $P = 0.5\,V^2$, $B = 10$, and $N = 2$, then $\sigma_o \leq 0.8\,\text{mV}$.

2.3.3   Impact of Gain

With the assumptions that the offset and timing skew for all N sub-ADCs are identical such that, without loss of generality, $o_i = 0$ and $\tau_i = 0$, the mean-square error in (2.39) reduces to

$$f(\hat{G}, \hat{\tau}) = \frac{P}{N} \sum_{i=0}^{N-1} G_i^2 - \frac{P}{N^2} \left(\sum_{i=0}^{N-1} G_i \right)^2 \tag{2.48}$$

Therefore, the *SNR* due to gain is

$$SNR_G = \frac{1}{\frac{1}{N} \sum_{i=0}^{N-1} G_i^2 - \frac{1}{N^2} \left(\sum_{i=0}^{N-1} G_i \right)^2} \tag{2.49}$$

Note that the *SNR* due to gain mismatch is independent of the signal power, and only depends on the magnitude of the individual gains. The statistical bound on the variance of gain, using (2.44), is

$$\sigma_G^2 \leq \left(\frac{N}{N-1} \right) \cdot \left(\frac{2}{3 \cdot 2^{2B}} \right) \tag{2.50}$$

This is almost identical to (2.47) in that it is inversely proportional to both the ADC resolution B and the interleaving factor N. However, it does not depend on the signal power P or on any other signal information. For example, if $N = 2$ and $B = 10$, then $\sigma_G \leq 1.1\%$.

2.3.4 Impact of Timing Skew

The results from analyzing timing skew are more interesting than those of both gain and offset, as they depend on the "speed" of the input signal. With the gain and offset of all N sub-ADCs set to $G_i = 1$ and offset $o_i = 0$, the mean-square error is

$$f(\hat{G}, \hat{\tau}) = P - \frac{1}{N^2 P} \left(\sum_{i=0}^{N-1} R(\tau_i - \hat{\tau}) \right)^2 \tag{2.51}$$

where

$$\hat{\tau} = \arg\max_{\tau} \sum_{i=0}^{N-1} R(\tau_i - \tau) \tag{2.52}$$

and thus the *SNR* is

$$SNR_\tau = \frac{1}{1 - \frac{1}{N^2} \left(\sum_{i=0}^{N-1} \frac{R(\tau_i - \hat{\tau})}{P} \right)^2} \tag{2.53}$$

The relationship between SNR_τ and the autocorrelation $R(\tau)$ in (2.53) is intuitive because the autocorrelation function reflects the "speed" of the signal. This is important since the speed, or the rate of change, of the input signal is directly proportional to the sampling error a given skew will create. For example, Fig. 2.10a shows a signal that does not change much for a certain value of τ, which leads to a small sampling error. This is captured by the autocorrelation $R(\tau)$, as in Fig. 2.10b, since $R(\tau)$ is close to 1. The signal in Fig. 2.10c changes significantly for a skew of τ. This is also captured by the autocorrelation $R(\tau)$, which, as in Fig. 2.10d, is not as close to 1.

A deterministic bound on timing skew is derived with

$$\frac{1}{P} \cdot \sum_{i=0}^{N-1} R(\tau_i - \hat{\tau}) \geq N \sqrt{\frac{SNR_Q - 1}{SNR_Q}} \tag{2.54}$$

Fig. 2.10 (a) Slow signal.
(b) Wide autocorrelation for
slow signal. (c) Fast signal.
(d) Narrow autocorrelation
for fast signal

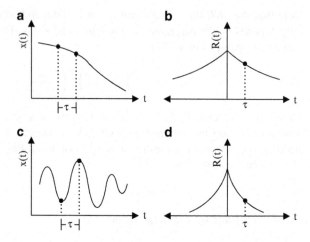

To calculate a statistical bound, it is useful to assume the autocorrelation is second-order differentiable, such that it can be expressed as a Taylor series centered around $\tau = 0$. Thus, when τ is small, we have

$$R(\tau) \approx R(0) + R'(0)\tau + \frac{R''(0)}{2}\tau^2 \tag{2.55}$$

where $R(0) = P$. Without loss of generality, $P = 1$. Since $R(\tau)$ is an even function and has a maximum at $\tau = 0$, $R'(0) = 0$ and $R''(0) \leq 0$. Therefore,

$$\frac{dR(\tau)}{d\tau} \approx R''(0)\tau \tag{2.56}$$

where $R''(0)$ is the curvature of the autocorrelation function.

Combining this with (2.41) allows us to solve for $\hat{\tau}$, such that

$$\sum_{i=0}^{N-1} R''(0)(\tau_i - \hat{\tau}) \approx 0$$

$$\hat{\tau} \approx \frac{1}{N} \sum_{i=0}^{N-1} \tau_i \tag{2.57}$$

Using (2.54) with (2.55) results in

$$\sum_{i=0}^{N-1} \left(1 + \frac{R''(0)}{2}(\tau_i - \hat{\tau})^2\right) \geq N\sqrt{\frac{SNR_Q - 1}{SNR_Q}} \tag{2.58}$$

and $R''(0)(\tau_i - \hat{\tau})^2$ can be expanded using (2.57) as

$$R''(0)(\tau_i - \hat{\tau})^2 = R''(0)(\tau_i^2 + \hat{\tau}^2 - 2\tau_i\hat{\tau})$$

$$= R''(0)\left(\tau_i^2 + \frac{1}{N^2}\left(\sum_{i=0}^{N-1}\tau_i\right)^2 - 2\frac{\tau_i}{N}\left(\sum_{i=0}^{N-1}\tau_i\right)\right) \quad (2.59)$$

Assuming the skews τ_i are independent and identically distributed random variables, with mean zero and variance σ_τ^2, the expected value of (2.58) is

$$E\left[\sum_{i=0}^{N-1} R(\tau_i - \hat{\tau})\right] \approx N + \frac{1}{2}R''(0)(N-1)\sigma_\tau^2 \quad (2.60)$$

and thus

$$N + \frac{1}{2}R''(0)(N-1)\sigma_\tau^2 \geq N\sqrt{\frac{SNR_Q - 1}{SNR_Q}} \quad (2.61)$$

Since

$$\sqrt{\frac{SNR_Q - 1}{SNR_Q}} \approx \left(1 - \frac{1}{2SNR_Q}\right) \quad (2.62)$$

for large SNR_Q, the variance σ_τ^2 is bounded by

$$\sigma_\tau^2 \leq \left(\frac{N}{N-1}\right) \cdot \left(\frac{2}{3 \cdot 2^{2B}}\right) \cdot \left(\frac{1}{|R''(0)|}\right) \quad (2.63)$$

This presents a closed-form bound on the acceptable variance of timing skew as a function of the number of sub-ADCs, the ADC resolution, and the curvature of the autocorrelation function, which is a property of the input signal statistics. Slower signals have smaller curvatures $R''(0)$, and thus have larger bounds.

Note that for a sinusoidal input with frequency f Hz, $R''(0) = -(2\pi f)^2$, and the bound on the variance $\hat{\sigma}_\tau^2$ using (2.63) is

$$\hat{\sigma}_\tau^2 \leq \left(\frac{N}{N-1}\right)\left(\frac{2}{3 \cdot 2^{2B}(2\pi f)^2}\right) \quad (2.64)$$

which matches that obtained in [12].

2.3.4.1 Wide-Sense Cyclostationary Signals

The above results for timing skew were obtained for WSS signals, but it is also possible to extend them to wide-sense cyclostationary (WSCS) signals. This is a more realistic model for some communication signals, such as those present in serial link receivers. A signal is WSCS if both its mean $m(t)$ and autocorrelation $R(t_1, t_2)$ are periodic in T [11], such that

$$m(t + T) = m(t) \tag{2.65}$$

$$R(t_1 + T, t_2 + T) = R(t_1, t_2) \tag{2.66}$$

For example, assume the autocorrelation of a zero-mean WSCS signal is periodic with the time-interleaved ADC sampling period T_s such that $R(t_1 + T_s, t_2 + T_s) = R(t_1, t_2)$. The ideal sampling phase of the first sub-ADC, which has previously been ignored because the input was WSS, is denoted by T_0, where $0 \le T_0 < T_s$, such that the autocorrelation changes depends on what T_0 is. Minimizing the mean-square error as done in previous sections and as elaborated on in Appendix A, respectively modifies (2.38) and (2.52) into

$$\hat{G} = \frac{\sum_i R(T_0 - \hat{\tau}, T_0 - \tau_i)}{N R(T_0 - \hat{\tau}, T_0 - \hat{\tau})} \tag{2.67}$$

and

$$\hat{\tau} = \arg\max_{\tau} \frac{\left(\sum_{i=0}^{N-1} R(T_0 - \hat{\tau}, T_0 - \tau_i) \right)^2}{N R(T_0 - \hat{\tau}, T_0 - \hat{\tau})} \tag{2.68}$$

The *SNR* in (2.53) and the variance in (2.63) then become a function of T_0.

2.3.4.2 Jitter

It is also possible to use (2.63) in bounding the tolerable random clock jitter for a single ADC or time-interleaved array by taking the limit of $N \to \infty$. This follows by noting that jitter causes the ADC to sample the signal with a different random phase τ_i for its ith sample, which is equivalent to having a time-interleaved ADC with an infinite number of sub-ADCs, such that each sub-ADC has timing skew τ_i and samples the input signal once. Thus, the bound on jitter is

$$\sigma^2 \le \left(\frac{2}{3 \cdot 2^{2B} |R''(0)|} \right) \tag{2.69}$$

This matches the result obtained by the authors of [13], who also show that using a sine wave in providing bounds on jitter overconstrains the variance bound by a factor of three when the input signal has a brick wall spectral density, as in (2.75). Equation (2.69) reduces to the known case of an input sine wave where $R''(0) = -(2\pi f)^2$ such that

$$\sigma^2 \le \left(\frac{2}{3 \cdot 2^{2B} (2\pi f)^2} \right) \tag{2.70}$$

Fig. 2.11 Setup for
simulation examples

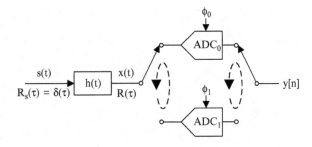

2.3.5 Simulation Examples

This section illustrates the preceding analysis on the effect of timing skew with examples of WSS wideband input signals, which are applied to both the deterministic and the statistical bounds. These signals are formed by coloring white noise with linear time-invariant filters, as in Fig. 2.11. The time-interleaved ADC used in these examples has $N = 2$ sub-ADCs. An example with a WSCS signal is also shown.

2.3.5.1 Examples for Deterministic Bounds

Both an ideal filter and a first-order low pass filter are used in this section, which allows us to compare the *SNR* obtained with (2.53) to that obtained with Monte Carlo simulations.

2.3.5.2 Ideal Filter

In this example, white noise is passed through an ideal low pass filter with cutoff frequency f_c Hz; the resulting signal has an autocorrelation function of

$$R(\tau) = \text{sinc}(2f_c\tau) \tag{2.71}$$

Without loss of generality, we set $\tau_0 = 0$, which is the timing skew of the first sub-ADC. This allows us to vary the timing skew τ_1 of the second sub-ADC and plot the theoretical value of (2.53) as a function of τ_1 for different values of f_c. This theoretical *SNR* is compared to that obtained with Monte Carlo simulations in Fig. 2.12 for different values of f_c. As is expected, the *SNR* increases for a given τ_1 as f_c decreases.

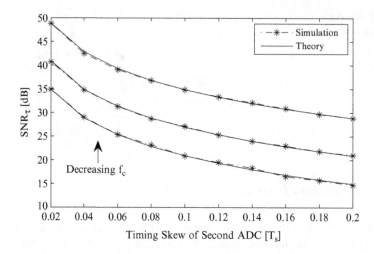

Fig. 2.12 Comparison of theoretical and simulation based SNR_τ with an input signal autocorrelation function of $R(\tau) = \mathrm{sinc}(2f_c\tau)$, for $f_c = 0.1f_s$, $0.25f_s$, and $0.5f_s$

2.3.5.3   First-Order Low Pass Filter

In this example, white noise is passed through a first-order low pass filter with a 3 dB frequency of f_{3dB} Hz. The autocorrelation of such an input signal is

$$R(\tau) = e^{-(2\pi f_{3dB})|\tau|} \tag{2.72}$$

The theoretical *SNR* obtained by replacing this in (2.53) is compared to the *SNR* obtained through Monte Carlo simulations in Fig. 2.13 for different values of f_{3dB}. Again, the achievable *SNR* depends on both the timing skew and f_{3dB}.

2.3.5.4   Examples for Statistical Bounds

This section demonstrates the applicability of (2.63) for WSS signals that have a second-order differentiable autocorrelation function. The examples used are the ideal filter and the second-order low pass filter.

2.3.5.5   Ideal Filter

Because $R(\tau)$ in this example [as in (2.71)] is second-order differentiable, we have

$$R''(0) = -\frac{1}{3}(2\pi f_c)^2 \tag{2.73}$$

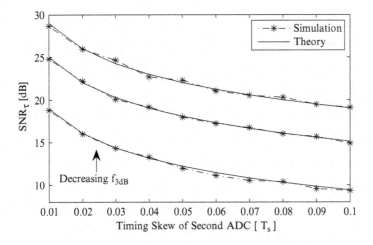

Fig. 2.13 Comparison of theoretical and simulation based SNR_τ with an input signal autocorrelation function of $R(\tau) = e^{-2\pi f_{3dB}|\tau|}$, for $f_{3dB} = 0.02 f_s$, $0.05 f_s$, and $0.2 f_s$

Replacing this in (2.63) results in a statistical bound of

$$\sigma_\tau^2 \leq \left(\frac{N}{N-1} \right) \cdot \left(\frac{2}{2^{2B} \cdot (2\pi f_c)^2} \right) \qquad (2.74)$$

Figure 2.14 shows how the ADC *SNR* is bounded by the sigma of the skew for different values of f_c as a function of f_s, the sampling frequency. It is worth looking at how the variance σ_τ^2 in (2.74) compares to that obtained with standard sinusoidal analysis. Since the input signal is bandlimited to f_c, standard analysis would use a sine wave of frequency f_c. The ratio of (2.74) to (2.64) is

$$\left(\frac{\sigma_\tau}{\hat{\sigma}_\tau} \right)^2 = 3 \qquad (2.75)$$

Thus, using standard sinusoidal analysis in this example leads to over-constraining the acceptable bound on timing skew variance by a factor of three, which also matches the result obtained in [13].

2.3.5.6 Second-Order Low Pass Filter

The impulse response for a second-order low pass filter is

$$h(t) = t e^{-(\omega_{3dB} t)} u(t) \qquad (2.76)$$

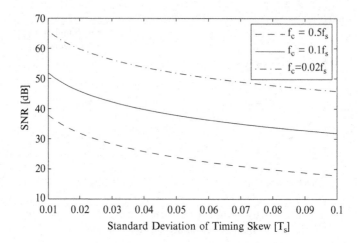

Fig. 2.14 ADC *SNR* as a function of the standard deviation of timing skew, which is calculated using equality in (2.74). Input signal is bandlimited white noise and has an autocorrelation function of $R(\tau) = \text{sinc}(2f_c\tau)$

where $\omega_{3\text{dB}} = 2\pi f_{3\text{dB}}$. The autocorrelation function for a second-order low pass filter, normalized such that $R(0) = 1$, is derived to be

$$R(\tau) = e^{-(|\tau|\omega_{3\text{dB}})} \cdot (1 + |\tau|\omega_{3\text{dB}}) \tag{2.77}$$

This is second-order differentiable, which allows us to calculate $R''(0)$ as

$$R''(0) = -(2\pi f_{3dB})^2 \tag{2.78}$$

Replacing this in (2.63) results in a statistical bound of

$$\sigma_\tau^2 \leq \left(\frac{N}{N-1}\right) \cdot \left(\frac{2}{3 \cdot 2^{2B} \cdot (2\pi f_{3\text{dB}})^2}\right) \tag{2.79}$$

A comparison to a sine wave is not as simple in this example as it is in the previous one, because the spectrum is nonzero for all frequencies. Therefore, assume that in the standard analysis, a sine wave with frequency $\hat{f}$ is used to calculate the bound on timing skew. This enables us to compare the bound on timing skew using (2.79) to that provided using (2.64) by setting $f_{3\text{dB}} = \alpha \hat{f}$. An interesting observation is that when $\alpha = 1$, or $f_{3\text{dB}} = \hat{f}$, the bound on skew is the same for both the second-order low pass filter and the sine wave input signal, even though the spectrum for the second-order low pass filter is still non-zero for frequencies larger than $\hat{f}$.

A more complete comparison is possible by looking at the ratio $\beta = \sigma_\tau/\hat{\sigma}_\tau$ as a function of α, as in Fig. 2.15, where σ_τ is defined in (2.79) and $\hat{\sigma}_\tau$ is defined in (2.64), such that

$$\beta = \frac{\sigma_\tau}{\hat{\sigma}_\tau} = \frac{1}{\alpha} \tag{2.80}$$

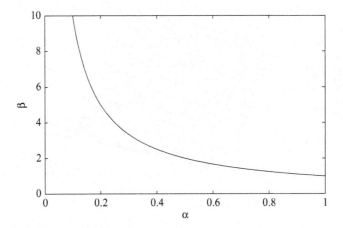

Fig. 2.15 Comparison of standard deviation of skew for second-order low pass filter and sine wave, where α is such that $f_{3dB} = \alpha \hat{f}$ and $\beta = \sigma_\tau / \hat{\sigma}_\tau$

For example, when $f_{3dB} = 0.5\hat{f}$, $\beta = 2$, which implies that standard analysis results in over-constraining the acceptable bound on the timing skew standard deviation by a factor of 2. In a more extreme example, when $f_{3dB} = 0.1\hat{f}$, $\beta = 10$.

This again demonstrates the importance of knowing the input signal statistics when deriving bounds on the acceptable timing skew. It is worth noting that even when f_{3dB} is not exactly known, as may be the case with certain signals, the range in which f_{3dB} falls can still be used. For example, if $0.1\hat{f} < f_{3dB} < 0.2\hat{f}$, then $10 > \beta > 5$.

2.3.5.7 Example of Deterministic Bound for WSCS Signals

In this example, an infinite series of bits $c_n \in \{-1, +1\}$, where $R_c(n, m) = E[c_n c_m] = \delta_{n-m}$ and $m_c = E[c_n] = 0$, are sent such that the transmitted signal is

$$s(t) = \sum_{m=-\infty}^{\lfloor t/T \rfloor} c_m p(t - mT) \tag{2.81}$$

where $p(t)$ is a rectangular pulse with length T and is defined by

$$p(t) = u(t) - u(t - T) \tag{2.82}$$

The transmitted signal s(t) passes through a linear time-invariant channel $h(t)$, as in Fig. 2.11, before the time-interleaved ADC can sample the signal $x(t)$. Thus,

a

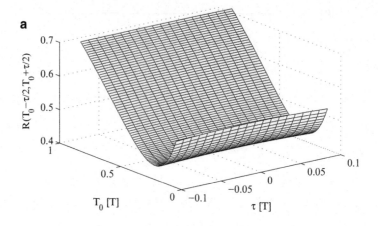

b

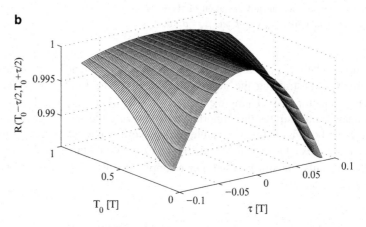

Fig. 2.16 Autocorrelation function $R(T_0 + \tau/2, T_0 - \tau/2)$ as a function of the sampling point T_0 and skew τ. Input signal is WSCS and has an autocorrelation function as in (A.11), with $\omega_{3dB} = 2/T$. (**a**) The actual autocorrelation function. (**b**) The autocorrelation function normalized such that $R(T_0, T_0) = 1$

$$x(t) = s(t) * h(t)$$

$$= \sum_{m=-\infty}^{\lfloor t/T \rfloor} c_m p(t - mT) * h(t)$$

$$= \sum_{m=-\infty}^{\lfloor t/T \rfloor} c_m f(t - mT) \qquad (2.83)$$

where $f(t)$ is the pulse response, defined by $f(t) = p(t) * h(t)$.

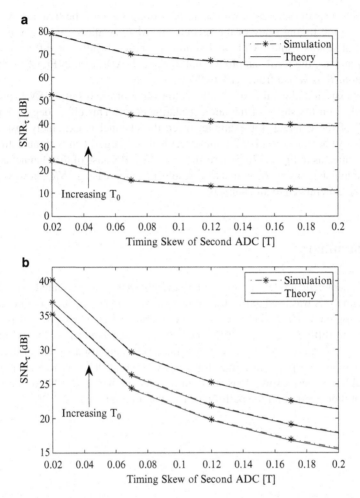

Fig. 2.17 Comparison of theoretical and simulation based SNR_τ. Input signal is WSCS and has an autocorrelation function as in (A.11). (**a**) With $\omega_{3dB} = 10/T$. (**b**) With $\omega_{3dB} = 1/T$

In this example, the channel $h(t)$ is a first-order low pass filter such that $h(t) = e^{-(t\omega_{3dB})}u(t)$. The autocorrelation function of $x(t)$ is

$$R(t_1, t_2) = E[x(t_1)x(t_2)] \tag{2.84}$$

and is fully derived in Appendix A, where it is also shown that $x(t)$ is WSCS.

An example of $R(t_1, t_2)$, where $t_1 = T_0 - \tau/2$ and $t_2 = T_0 + \tau/2$, is shown in Fig. 2.16a as a function of T_0 and the skew τ for $\omega_{3dB} = 2/T$. Figure 2.16b uses a normalized version of $R(t_1, t_2)$ such that $R(T_0, T_0) = 1$, which illustrates the change in the curvature of the autocorrelation function as a function of T_0.

Without loss of generality, we set τ_0, the timing skew of the first sub-ADC, to 0. Varying the timing skew of the second sub-ADC τ_1 allows us to compare the theoretical results using (A.11) and simulation based results for different values of ω_{3dB} and T_0, the ideal sampling point of the sub-ADCs, as in Fig. 2.17. In this simulation, T_0 is varied from $0.1T$ to $0.7T$.

As is expected, the value of T_0 affects the value of the resulting SNR_τ because of its effect on the shape of the autocorrelation curve. This effect depends on the "speed" of the channel; for example, when the channel is extremely fast, as in Fig. 2.17a (where $\omega_{3dB} = 10/T$), the effect is much larger than with an extremely slow channel, as in Fig. 2.17b (where $\omega_{3dB} = 1/T$). Because of the channel used in this example, increasing T_0 from 0 to T results in an increasing SNR_τ; however, this cannot be generalized to all channels.

2.4 Summary

In this chapter, a model for time-interleaved ADCs was presented. Frequency domain analysis was used to illustrate how time-varying errors, such as gain, offset, and timing skew, affect the resulting time-interleaved ADC output. Expressions relating the different errors to ADC performance and bounds on the magnitude of these errors were also derived, and simulations were used to demonstrate the accuracy of these expressions. Thus, for the given set of ADC specifications required by serial links, these expressions are used to calculate the acceptable timing skew, such that it does not limit the performance of the time-interleaved ADC.

Chapter 3
Mitigation of Timing Skew

Time-varying errors degrade the performance of time-interleaved ADCs, as discussed in Chap. 2. Since the effect of timing skew increases with input frequency, it overshadows the effect of gain and offset when input signals with multi-GHz frequencies are sampled. As the input signal frequencies increase, the constraint on timing skew grows more stringent and can reach the sub-picosecond range. Designing a time-interleaved ADC to meet yield constraints on timing skew without extra correction circuitry is possible [14] only for limited timing skew bounds, primarily because the residual timing skew is generally not within the designer's control. This chapter discusses timing skew and its sources in more detail, and describes and analyzes the use of a statistics-based background calibration algorithm to mitigate the impact of timing skew.

3.1 Bounds on Timing Skew

As derived in Chap. 2, the statistical bound on timing skew is a function of the input signal statistics. For input sinusoidal signals with a frequency of f_{in}, the bound was shown to be

$$\sigma_\tau \leq \sqrt{\left(\frac{N}{N-1}\right) \cdot \left(\frac{2}{3 \cdot 2^{2B}}\right) \cdot \left(\frac{1}{(2\pi f_{in})^2}\right)} \tag{3.1}$$

where N is the interleaving factor and B is the ADC resolution. In Fig. 3.1, the ADC resolution B is plotted as a function of the standard deviation σ_τ for different input frequencies, and the horizontal line denotes the maximum acceptable standard deviation of timing skew to achieve a 5 bit resolution. For signals with a frequency greater than 4 GHz, sub-picosecond timing skew is required if this resolution is to be obtained.

M. El-Chammas and B. Murmann, *Background Calibration of Time-Interleaved Data Converters*, Analog Circuits and Signal Processing, DOI 10.1007/978-1-4614-1511-4_3, © Springer Science+Business Media, LLC 2012

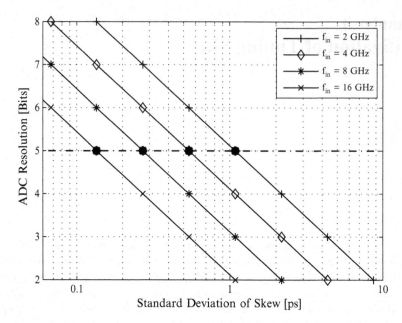

Fig. 3.1 Bounds on the ADC resolution

3.2 Sources of Timing Skew

Ideally, the N sub-ADCs consecutively sample the input signal at times nT_s, where T_s is the sampling period of the time-interleaved ADC. This is achieved by having the sampling points of two consecutive sub-ADCs separated by a timing offset of T_s, where each sub-ADC has a clock period of $\hat{T}_s = N \cdot T_s$. A phase generator creates the sub-ADC clocks $\phi_i(t)$, as in Fig. 3.2a, which ideally have sampling edges spaced as in Fig. 3.2b. However, two types of circuit mismatch affect both the signal and clock propagation delay, resulting in non-zero timing skew, and prevent the uniform sampling of the input signal. The first is due to transistor variations and the second to trace and load variations; both sources of timing skew are discussed in this section.

3.2.1 Transistor Variations

The outputs of the phase generator are followed by a series of buffers, as shown in Fig. 3.2a, which then drive the clocking distribution network for each sub-ADC. Due to random variations, such as those in transistor threshold voltages [15], the buffer delays vary and timing skew results.

The threshold variations are inversely proportional to the transistor area [15]. Decreasing the variations by sizing up the transistors leads to both improved timing skew and an increase in power. Computer simulations using TSMC 65 nm GP models were run to show this relationship.

Fig. 3.2 (a) Sub-ADC clocks created by phase generator. (b) Sampling edges of sub-ADC clocks

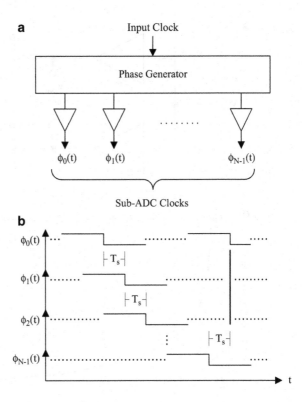

For example, in Fig. 3.3a, the outputs V_{out1} and V_{out2} of the FO4-sized inverters should ideally be identical, since both inverter chains have the same input. However, due to variations, this is not the case. It is possible to plot the resulting standard deviation of timing skew between the two outputs V_{out1} and V_{out2} as a function of the required power by running Monte Carlo simulations while increasing the size of the inverters. As is clear from Fig. 3.3b, to reduce the timing skew due to threshold variations, more and more power must be invested in the inverters, with diminishing returns. As an approximate rule of thumb, reducing the timing skew by a factor of two requires four times as much power.

3.2.2 Trace and Load Variations

A second source of timing skew is that of trace and load variations. Trace variations arise from nonuniformity in interconnect widths and thicknesses, which affect the trace resistance and capacitance and thus alter the propagation delay. Load variations are due to changes in the input load of the following stage. These are problematic as the delay of an inverter is proportional to its load. For example, if a

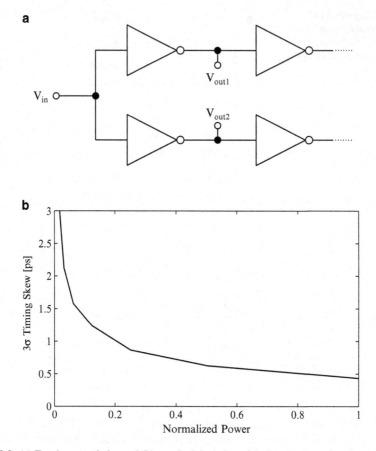

Fig. 3.3 (a) Two inverter chains and (b) standard deviation of timing skew as a function of power

two inverter chains, as in Fig. 3.4a, are simulated, such that the load capacitance of each stage is slightly different, then it is possible to plot the effect on timing skew as a function of load variations, as in Fig. 3.4b.

3.2.3   Cumulative Effects of Variations

The examples in Figs. 3.3a and 3.4a each deal with the effects of variations on one inverter delay. In reality, a clock distribution circuit consists of a phase generator, which could be either a PLL or a DLL, output buffers for each of the phases, sampling switches, and a mixture of interconnects and vias, as in Fig. 3.5. Each of these elements suffer from threshold and load variations. These effects accumulate and can result in more than 10 ps of timing skew [16]. With high-speed input signals, such timing skew is detrimental, as seen in Fig. 3.1.

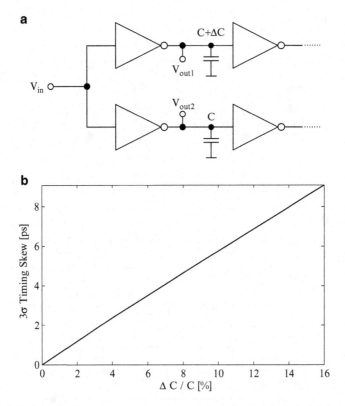

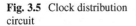

Fig. 3.4 (**a**) Two inverter chains with load variations and (**b**) standard deviation of timing skew as a function of load variations

Fig. 3.5 Clock distribution circuit

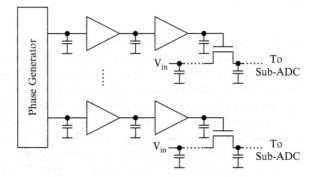

3.3 Timing Skew Mitigation

It is possible to make the time-interleaved ADC insensitive to the effect of timing skew by using a single track-and-hold in front of all the sub-ADCs [17, 18], as in Fig. 3.6a. The clock $\phi_{TH}(t)$ has the same frequency as the sample rate of the

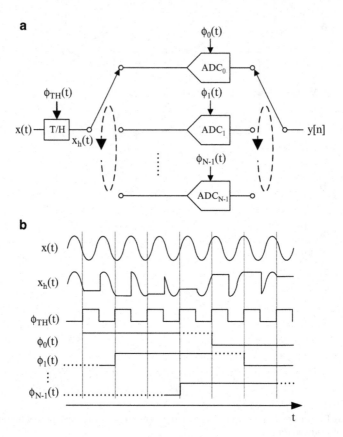

Fig. 3.6 (**a**) Single sampler used for all sub-ADCs. (**b**) Signal and clock waveforms when single sampler is used

time-interleaved ADC. This creates $x_h(t)$, which has a constant value when the switch is open, as in Fig. 3.6b. Thus, the sub-ADC samples a constant voltage and can accept some timing skew in its sampling point. Unfortunately, this solution is not practical in multi-GS/s designs, due to limitations with the track-and-hold.

Another approach to mitigating the effect of timing skew is to correct it. There are two main techniques for compensating the effects of timing skew, which can be extended to other time-varying errors. The first operates in the digital domain by appending a digital processor to the outputs of the sub-ADCs [19], such that the processor corrects the digital outputs. Figure 3.7a displays the case of two interleaved sub-ADCs with a digital processor. The outputs of the sub-ADCs $y_0[n]$ and $y_1[n]$ each pass through an adaptive filter that corrects for the effect of timing skew such that the combination of the two digitally corrected outputs $\tilde{y}_0[n]$ and $\tilde{y}_1[n]$ does not suffer degradation. The adaptive filters are tuned with a detection block that can run various algorithms on the sub-ADC outputs.

This technique requires the use of fractional delay filters [20], which interpolate between the sub-ADC samples to overcome the effects of timing skew. However,

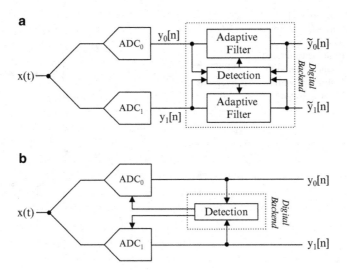

Fig. 3.7 Correction in the (**a**) digital domain and (**b**) mixed-signal domain

the nature of the fractional delay filter leads to a high complexity in the number of filter taps required, and the power consumption of the digital correction system is a limiting barrier when it comes to implementing such an architecture. Although this approach may be tractable for lower frequency designs, multi-GS/s ADCs suffer a large power penalty that currently makes this infeasible in serial links.

The second technique for compensating the effects of timing skew operates in the mixed-signal domain by using a digital backend to detect certain characteristics of the discrete-time output and to then adjust analog circuits in order to compensate for the effect of timing skew [21], as in Fig. 3.7b. This approach increases the design space and can potentially lead to the power-efficient partitioning of tasks between the analog and digital domain.

Timing skew correction can be accomplished by using either foreground or background calibration. Foreground calibration, as shown in Fig. 3.8a,b, separates the parameter calibration process and the operation of the ADC. In order for the ADC to be calibrated, it must be taken offline, in which case it samples a test signal, and not the actual input signal. When the ADC is placed back online in normal operation, it can no longer be calibrated. Foreground calibration has its applications, and may be used when circuit parameters do not vary much with environmental changes, such as those of voltage or temperature, or when the application allows the ADC to be intermittently taken offline for calibration, such as in oscilloscopes [21].

In applications where circuit parameters do vary or where disconnecting the ADC is not an option, such as in communication links, foreground calibration is not a practical solution. Background calibration is much more attractive, and, as in Fig. 3.8c, enables the ADC to be calibrated during normal operation, and thus allows the ADC to process the input while the calibration algorithm tracks environmental changes.

Fig. 3.8 With foreground
calibration, (**a**) ADC is either
online and samples input or
(**b**) is offline and is calibrated.
With background calibration,
(**c**) ADC is calibrated while
sampling the input signal

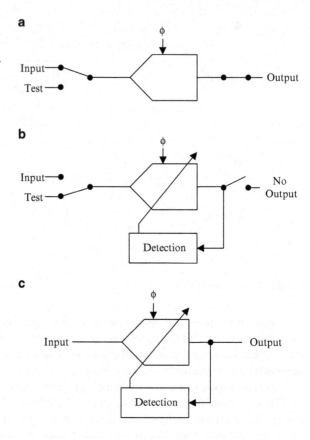

Most background calibration techniques published to date suffer from various
signal constraints [22], which are not always guaranteed in wireline systems. The
method presented in the remainder of this chapter greatly relaxes the input signal
bandwidth constraints, and results in a solution that has only a marginal power
increase.

3.4   Background Timing Skew Calibration

Compensating for timing skew, regardless of whether a digital or mixed-signal
approach is used, improves the performance of the time-interleaved ADC. The
relationship between timing skew and *SNR* was derived in Chap. 2, as shown in
(2.53). Thus, maximizing the *SNR* is equivalent to [23]

$$\sum_{i=1}^{N-1} \left(\max_{\tau_i} R(\tau_i) \right) \tag{3.2}$$

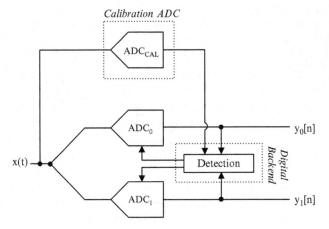

Fig. 3.9 Attaching a calibration ADC to the time-interleaved array

where $R(\tau)$ is the autocorrelation of the input signal. The maximum of (3.2) occurs at $\tau_i = 0$ for all i, which is intuitive as all the timing skews have been minimized.

Unfortunately, calculating the autocorrelation of the input signal using the sub-ADC outputs is not possible. However, the input autocorrelation can be replaced by the crosscorrelation between the outputs of each sub-ADC and an additional sub-ADC, as shown in Fig. 3.9, which also has a maximum at $\tau_i = 0$. Thus, implementing (3.2) occurs in two steps. The first step is to calculate the crosscorrelation for each sub-ADC, and the second step is to maximize it by adjusting the value of τ_i. This is iteratively implemented for each sub-ADC until τ_i converges to zero for all i, which achieves the main goal of maximizing the *SNR*.

3.4.1 Calculating the Correlation

As in Fig. 3.9, an additional ADC is used to calculate the crosscorrelation for each sub-ADC output. Thus, if N sub-ADCs are being interleaved, the overall ADC has a total of $N + 1$ sub-ADCs. The extra ADC does not contribute to the output of the time-interleaved ADC; it only feeds information to the digital calibration backend, which calculates the crosscorrelation between each sub-ADC and the calibration ADC. This is further elaborated on by focusing on a single sub-ADC and the calibration ADC and momentarily assuming both the sub-ADC and the calibration ADC have the same sample rate $\hat{f}_s$. The timing skew between the sampling points of the two sub-ADCs is τ.

Ignoring quantization effects, the digital backend calculates $\hat{R}(\tau)$, an approximate version of the crosscorrelation $R(\tau)$, by multiplying the outputs of the sub-ADC and calibration ADC, $y[n]$ and $y_c[n]$, respectively, where $y[n] = x(nT_s - \tau)$ and $y_c[n] = x(nT_s)$, and averaging this product over M samples, as in Fig. 3.10a. Therefore,

Fig. 3.10 (a) Calculating the correlation between the calibration ADC and the sub-ADC. (b) Maximizing the correlation with a variable delay line

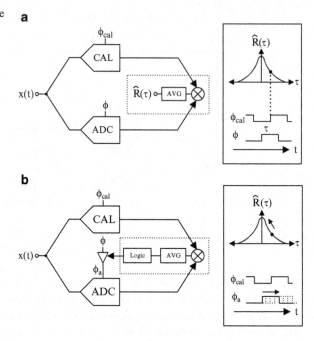

$$\hat{R}(\tau) = \frac{1}{M} \sum_{n=1}^{M} y[n] y_c[n]$$

$$= R(\tau) + E(M) \tag{3.3}$$

where $E(M)$ is the error term between the crosscorrelation $R(\tau)$ and its approximation $\hat{R}(\tau)$. The variance of $E(M)$ is inversely proportional to M [24].

3.4.2   Maximizing the Correlation

The background algorithm maximizes the crosscorrelation by adjusting the value of τ, which is the timing skew between the sampling points of the sub-ADC and the calibration ADC. This is achieved by adding a variable delay line that closes the calibration loop, as in Fig. 3.10b, such that the delay line adjusts the sampling edge of the sub-ADC in a direction that maximizes $R(\tau)$.

3.4.3   Simplifying the Algorithm

Two simplifications can be made to the calibration algorithm. The first is to reduce the resolution of the calibration ADC and the second is to decrease the sampling rate of the calibration ADC.

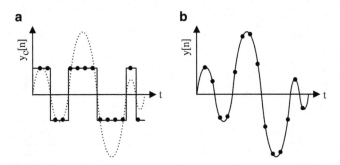

Fig. 3.11 (**a**) Output of single-bit calibration ADC. (**b**) Output of sub-ADC

3.4.3.1 Reducing the Resolution of the Calibration ADC

The calibration algorithm does not require the calibration ADC and sub-ADC to have the same transfer function. Hence, it is possible to reduce the resolution of the calibration ADC to a single bit, such that the outputs of the calibration ADC and the sub-ADC are as shown in Fig. 3.11. Reducing the resolution of the calibration ADC does not change the shape of the correlation function. If $R(\tau)$ is the autocorrelation of a signal $x(t)$, the correlation between $x(t)$ and some nonlinear function of $x(t)$ is simply a scaled version of $R(\tau)$ [25], so that the resolution of the calibration ADC can indeed be reduced without loss of detail in the correlation function.

This can be taken further by calculating the correlation using only a one-bit representation of the sub-ADC, similar to that in [26], in addition to a one-bit representation of the calibration ADC. The resulting correlation is

$$R_1(\tau) = \frac{2}{\pi} \sin^{-1}(R(\tau)) \tag{3.4}$$

which is known as the Van Vleck relationship [27]. Unfortunately, larger quantization in the sub-ADC renders this approach more susceptible to ADC offsets. This can be demonstrated by taking an input sinusoidal function with frequency f_{in} and unit amplitude. Without loss of generality, allowing only the calibration ADC to have an offset of $v_o \geq 0$ results in a correlation function of

$$R(\tau) = \begin{cases} 1 - \frac{2}{\pi}\sin^{-1}(v_o) & \text{if } |\tau| \leq \frac{1}{2\pi f_{in}}\sin^{-1}(v_o) \\ 1 - 4f_{in}\tau & \text{if } \frac{1}{2\pi f_{in}}\sin^{-1}(v_o) < |\tau| \leq \frac{1}{2f_{in}} - \frac{1}{2\pi f_{in}}\sin^{-1}(v_o) \\ \frac{2}{\pi}\sin^{-1}(v_o) - 1 & \text{if } \frac{1}{2f_{in}} - \frac{1}{2\pi f_{in}}\sin^{-1}(v_o) < |\tau| \leq \frac{1}{2f_{in}} \end{cases} \tag{3.5}$$

When $|\tau| \leq \frac{1}{2\pi f_{in}}\sin^{-1}(v_o)$ in (3.5), the autocorrelation is flat, as in Fig. 3.12, and thus there is not a unique maximum that the calibration algorithm can converge to. This is not a problem if the flat region is smaller than the bound on skew, as in (2.64), which leads to a bound on the acceptable offset v_o. The bound on the standard

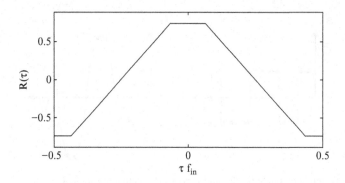

Fig. 3.12 Correlation of single-bit outputs with offset

deviation of v_o that reduces the flat region in (3.5) to within the timing skew bound is

$$\sigma_{v_o} \leq \sin(2\pi f_{in}\sigma_\tau) \approx 2\pi f_{in}\sigma_\tau$$

$$\leq \sqrt{\left(\frac{N}{N-1}\right) \cdot \left(\frac{2}{3 \cdot 2^{2B}}\right)} \tag{3.6}$$

under the assumption that $2\pi f_{in}\sigma_\tau \ll 1$.

When more than one bit is used from the sub-ADC, more offset is acceptable, as it translates into a smaller flat region, and thus (3.6) provides a pessimistic upper bound. However, offset correction for the calibration ADC may still be required to achieve the necessary time resolution.

3.4.3.2 Decreasing the Sampling Frequency of the Calibration ADC

The second simplification to the algorithm is to reduce the sample rate of the calibration ADC, as long as, for large K,

$$R(\tau) \approx \frac{1}{K} \cdot \sum_{i=0}^{K-1} y_c(iT_1) \cdot y(iT_1)$$

$$\approx \frac{1}{K} \cdot \sum_{i=0}^{K-1} y_c(iT_2) \cdot y(iT_2) \tag{3.7}$$

where $y_c(t)$ is the output of the calibration ADC, $y(t)$ the output of the sub-ADC, and $T_1 \neq T_2$. A sufficient, but not necessary, condition is for the input signal to be ergodic [24]. This allows the correlation be calculated with a slower calibration clock frequency, which leads to a decrease in power in both the calibration ADC and the digital backend. It also has the more important benefit of allowing the calibration

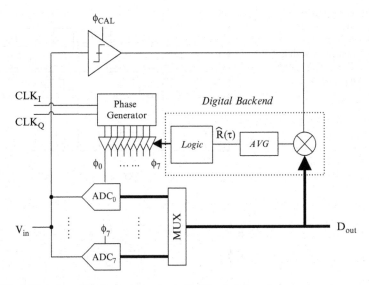

Fig. 3.13 Adding the calibration comparator to the time-interleaved array

ADC to cycle through all the sub-ADCs, as discussed in the following section, as it does not need to sample the signal at the same rate as the sub-ADCs.

3.4.4 Calibrating All the Sub-ADCs

In the previous discussion, the calibration ADC was used with a single sub-ADC. This is extendable to the time-interleaved ADC by adding a single calibration ADC, which is implemented with a single comparator, and providing each sub-ADC with a delay line, as in Fig. 3.13. By using the calibration ADC as a timing reference and creating a timing grid that matches the ideal sampling points of all the sub-ADCs, it is possible to minimize the timing differences between all the sub-ADCs and the calibration ADC.

This is accomplished by controlling the calibration ADC with a clock such that the sampling edge of the calibration clock cycles through the ideal sampling points of the sub-ADC clocks. Thus, as seen in Fig. 3.14, the first sampling edge of the calibration clock coincides with the ideal sampling point of the first sub-ADC, which allows the digital backend to calculate the crosscorrelation for the first sub-ADC. The second sampling edge coincides with that of the second sub-ADC, such that the crosscorrelation of the second sub-ADC is calculated, and so forth, for all the sub-ADCs. In general, a calibration clock frequency of $\frac{f_s}{M}$, where f_s is the sample rate of the time-interleaved ADC and where the greatest common denominator of M and the interleaving factor N is one, is sufficient.

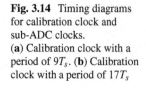

Fig. 3.14 Timing diagrams for calibration clock and sub-ADC clocks.
(**a**) Calibration clock with a period of $9T_s$. (**b**) Calibration clock with a period of $17T_s$

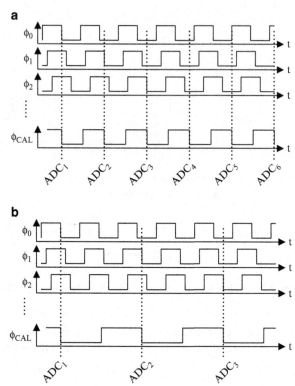

Figure 3.14 shows two clock timing diagrams for a time-interleaved ADC with eight sub-ADCs, where Fig. 3.14a,b have a calibration clock frequency of $f_s/9$ and $f_s/17$, respectively.

3.4.4.1   Clocking the Calibration ADC

In the prototype ADC described in Chap. 5, an external signal generator was used for the calibration clock. However, depending on the relationship between the reference clock frequency and the time-interleaved ADC sampling frequency, two alternate approaches can be used in SoC environments.

For lower frequency ADCs, it is possible to provide a reference clock that has the same frequency as the sampling rate [14]. In this scenario, the calibration clock can be created by using a control block to clock-gate the reference clock and to select the required sampling edges, as in Fig. 3.15a. An important feature to keep in mind in this approach is that the calibration clock path must always be constant, such that the calibration clock passes through the same mismatches. Periodic changes in the clock path create harmonics, which translate into deterministic skew since the timing reference provided by the calibration ADC no longer matches the ideal sampling points of all the sub-ADCs.

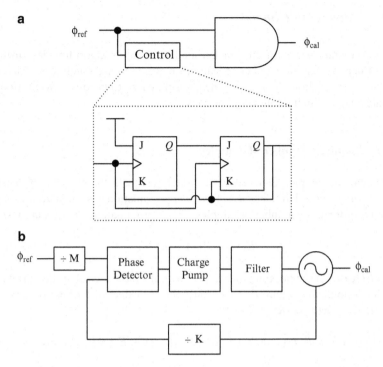

Fig. 3.15 (**a**) Clock-gating to create the calibration clock, with an example control circuit for divide-by-three. (**b**) Using an integer-PLL to create the calibration clock

When the reference clock has a frequency of $\tilde{f} = \frac{f_s}{K}$, then the frequency of the calibration clock can be $f_{cal} = \frac{K \cdot \tilde{f}}{M} = \frac{f_s}{M}$, where the greatest common denominator of the interleaving factor N and M is one. This is sufficient for the calibration ADC to cycle through the sub-ADCs, and such a clock can be created by either using an integer-PLL or a fractional-PLL. For example, an integer-PLL would initially divide the reference clock by M, and then multiply it by K by having a divide-by-K counter in the feedback loop, as in Fig. 3.15b.

3.5 Algorithmic Behavior

Section 3.4 presented a statistics-based background calibration algorithm. The convergence speed of a calibration algorithm is an important feature and is discussed in this section. Furthermore, conditions on the input signal ensuring that the proposed algorithm works, and the effect of quantization, which was previously ignored, are both discussed.

3.5.1 Convergence Speed

The convergence speed of the background calibration algorithm described in Sect. 3.4 depends on the number of samples required to accurately calculate each value of the correlation $R(\tau)$ and on the algorithm used to maximize $R(\tau)$. Both of these are analyzed in the following sections.

3.5.1.1 Required Number of Samples

The number of samples required to accurately estimate the correlation depends primarily on the correlation curve, since the approximation has a non-zero variance as a result of the finite number of samples. The approximation $\hat{R}(\tau)$, as in (3.3), is

$$\hat{R}(\tau) = R(\tau) + E(N) \tag{3.8}$$

where $E(N)$ is the approximation noise. Since it is assumed that the correlation curve has a single maximum, $R(\tau_1) < R(\tau_2)$ for two values of τ where $\tau_1 > \tau_2 > 0$. The difference between $\hat{R}(\tau_1)$ and $\hat{R}(\tau_2)$ is

$$F(\tau_1, \tau_2) = \hat{R}(\tau_2) - \hat{R}(\tau_1) = R(\tau_2) - R(\tau_1) + E_2(N) - E_1(N) \tag{3.9}$$

In (3.9), $F(\tau_1, \tau_2)$ is not guaranteed to be larger than 0 because of the residual error terms, even though $R(\tau_2) - R(\tau_1) > 0$. The probability that it is larger than 0 is a function of the distribution of $F(\tau_1, \tau_2)$, which has a mean of $R(\tau_2) - R(\tau_1)$. This is a typical problem with such averaging systems and in the specific context of estimating correlations. A larger change in the correlation for a given τ_2 and τ_1, which corresponds to a "fast" signal, results in a higher probability that $F(\tau_1, \tau_2) > 0$ than a smaller change in the correlation.

The usual conclusion to draw from this is that more samples are needed for slower signals, which can be illustrated with an example. Assume a sinusoidal input signal with frequency f such that $x(f,t) = 2\sin(2\pi f t)$ and $R(f, \tau) = \cos(2\pi f t)$. Given τ and $\Delta\tau$ such that $|2\pi f \tau| \ll 1$ and $|\Delta\tau| \ll 1$, the difference between $R(f, \tau)$ and $R(f, \tau + \Delta\tau)$ is

$$\Delta R(f, \tau) = R(f, \tau + \Delta\tau) - R(f, \tau) \approx \frac{dR(f, \tau)}{d\tau} \Delta\tau \tag{3.10}$$

such that with two signals $x(f_1, t)$ and $x(f_2, t)$, where $f_2 = c \cdot f_1$ for $c > 0$,

$$\Delta R(f_1, \tau) \approx -2\pi f_1 \sin(2\pi f_1 \tau)\Delta\tau \approx -(2\pi f_1)^2 \tau \Delta\tau$$
$$\Delta R(f_2, \tau) \approx -2\pi f_2 \sin(2\pi f_2 \tau)\Delta\tau \approx -(2\pi f_2)^2 \tau \Delta\tau \tag{3.11}$$

With a finite number of samples, the variance for $\Delta R(f_1, \tau)$ and $\Delta R(f_2, \tau)$ is σ_1^2 and σ_2^2, respectively. The number of required samples is set by collecting enough data such that

$$\Delta R(f_1, \tau) = k\sigma_1$$
$$\Delta R(f_2, \tau) = k\sigma_2 \tag{3.12}$$

for some value of k. Thus,

$$\frac{\Delta R_1}{\Delta R_2} = \frac{\sigma_1}{\sigma_2} = \sqrt{\frac{N_2}{N_1}} \tag{3.13}$$

since σ_1 and σ_2 are inversely proportional to $\sqrt{N_1}$ and $\sqrt{N_2}$, respectively. Since $f_2 = c \cdot f_1$

$$\sqrt{\frac{N_2}{N_1}} \approx \frac{-(2\pi f_1)^2 \tau \Delta \tau}{-(2\pi c \cdot f_1)^2 \tau \Delta \tau} = \frac{1}{c^2} \tag{3.14}$$

and

$$N_1 \approx c^4 N_2 \tag{3.15}$$

which implies that the number of samples varies with the 4th power of the frequency ratio, c. For example, if $f_2 = 0.5 f_1$, then the calibration will need 16 times as many samples in order to obtain a similar approximation in terms of accuracy, for the same step $\Delta \tau$.

Fortunately, this argument is overly pessimistic in the context of timing skew for time-interleaved ADCs, since the bound on timing skew is a function of the input frequency, as derived in Chap. 2. For a sinusoidal input, the bound on timing skew is

$$\sigma_\tau \leq \sqrt{\left(\frac{N}{N-1}\right) \cdot \left(\frac{2}{3 \cdot 2^{2B}}\right) \cdot \left(\frac{1}{(2\pi f_{in})^2}\right)} \tag{3.16}$$

The change $\Delta \tau$ is comparable to σ_τ, and thus is a function of the input frequency. Equation (3.11) becomes

$$\Delta R(f_1, \tau) \approx -2\pi f_1 \sin(2\pi f_1 \tau_1) \Delta \tau \approx -(2\pi f_1)^2 \tau_1 \Delta \tau_1$$
$$\Delta R(f_2, \tau) \approx -2\pi f_2 \sin(2\pi f_2 \tau_2) \Delta \tau \approx -(2\pi f_2)^2 \tau_2 \Delta \tau_2 \tag{3.17}$$

where both τ_1 and τ_2 are chosen such that $x(f_1, \tau_1) = x(f_2, \tau_2)$, as this ensures similar ADC performance. Thus,

$$\sqrt{\frac{N_2}{N_1}} = \frac{f_1 \Delta \tau_1}{f_2 \Delta \tau_2} = \frac{1}{c} \cdot \frac{\sqrt{\left(\frac{N}{N-1}\right) \cdot \left(\frac{2}{3 \cdot 2^{2B}}\right) \cdot \left(\frac{1}{(2\pi f_1)^2}\right)}}{\sqrt{\left(\frac{N}{N-1}\right) \cdot \left(\frac{2}{3 \cdot 2^{2B}}\right) \cdot \left(\frac{1}{(2\pi c f_1)^2}\right)}} = 1 \tag{3.18}$$

The number of points required to achieve similar accuracy is the same, because the timing resolution required for lower frequencies is larger due to the relaxed bounds on timing skew.

3.5.1.2   Digital Algorithm

The aim of the calibration algorithm is to maximize the correlation. However, its implementation affects the speed of convergence and the complexity of the digital backend. Two algorithms are presented below; the first is a simple iterative maximizer and the second is a gradient based stochastic maximizer.

The calibration algorithm discretely adjusts the timing skew of the sub-ADCs with a digitally controlled delay line. The algorithm implementation is divided into calibration cycles, such that each calibration cycle consists of N samples for each sub-ADC. At the end of the nth calibration cycle, a correlation value of $\hat{R}[n]$ is estimated using a cumulative adder, which corresponds to a skew correction code of $D[n]$. Based on this correlation value and previous history, the skew correction code $D[n + 1]$ is set, and the next calibration cycle begins.

The iterative maximizer adjusts the skew correction code, which is the digital input to the delay line, by incrementing or decrementing the code by a single bit. Thus, if the algorithm detects that the delay must be increased in order to approach the maximum of the correlation, the delay code $D[n]$ is adjusted such that $D[n + 1] = D[n] + 1$. This results in a simple algorithm only consisting of a series of digital adders and comparators.

The change in $D[n + 1]$, where

$$D[n + 1] = D[n] \pm 1 \tag{3.19}$$

depends on the outputs of the two comparisons

$$\hat{R}[n] \gtrless \hat{R}[n - 1]$$
$$D[n] \gtrless D[n - 1] \tag{3.20}$$

Thus, if

$$A = \text{sign}\left(\hat{R}[n] - \hat{R}[n - 1] \right) \tag{3.21}$$

and

$$B = \text{sign}(D[n] - D[n - 1]) \tag{3.22}$$

then

$$D[n + 1] = D[n] + A \cdot B \tag{3.23}$$

which is an easily implemented update formula. The main drawback in this approach is that the comparisons have binary outputs and dispense with the correlation differences, which may contain valuable information that can be used to improve convergence.

The gradient based stochastic maximizer is an algorithm that makes use of gradient information to adjust the value of $D[n]$ and speed up convergence. This is an LMS-based algorithm that updates $D[n]$ to converge to $R'(\tau) = 0$. For curves with a continuous derivative, this is equivalent to $\tau = 0$. $D[n]$ is updated with

$$D[n + 1] = D[n] + \mu \hat{R}'[n] \tag{3.24}$$

where μ is the step size. The gradient is approximated with

$$\hat{R}'[n] = \frac{\hat{R}[n] - \hat{R}[n - 1]}{D[n] - D[n - 1]} \tag{3.25}$$

which is related to the gradient $R'(0)$ with

$$\hat{R}'[n] = R'[n] + e[n] \tag{3.26}$$

Thus,

$$D[n + 1] = D[n] + \mu(R'[n] + e[n])$$
$$= (D[n] + \mu R'[n]) + \mu e[n] \tag{3.27}$$

The noise in the updated skew correction code $D[n + 1]$ has a variance proportional to μ^2/N. Thus, decreasing the value of μ, which will result in smaller updates, also allows the reduction of N, the number of samples in each calibration cycle.

This approach, which works for a smaller set of signals than the first approach, allows the dynamic throttling of μ. In the startup stages, μ can be increased while still maintaining stability, in order to speed up the convergence. Once convergence has been achieved, μ can be decreased to reduce the variance on the update noise.

3.5.2 Conditions on Input Signal

In order for the calibration algorithm to work, the input signal $x(t)$ must have signal activity around the calibration ADC trip point, which in this implementation is $x(t) = 0$. Furthermore, some stationarity conditions on the input signal are required, since the calibration algorithm estimates the correlation over a period of time and compares it to previous correlation values. Since all that is required for the algorithm is the value of the correlation, wide-sense stationarity requirements are sufficient. This can be relaxed if the signal is sample-invariant, such that

$$\lim_{N \to \infty} \frac{1}{N} \sum_{n=m_1}^{N+m_1} x(nT_s) \cdot x(nT_s - \tau) = \lim_{N \to \infty} \frac{1}{N} \sum_{n=m_2}^{N+m_2} x(nT_s) \cdot x(nT_s - \tau) \tag{3.28}$$

where $x(t)$ is the input signal and where $m_1 \neq m_2$. However, it can still be acceptable if (3.28) is not true, as long as the autocorrelation changes "slowly," where "slowly" means compared to the calibration algorithm convergence speed.

The other condition on the input signal is that the autocorrelation of the input signal must have a single maximum within the region of concern. This is defined as the expected skew the ADC suffers from. For example, if the sub-ADCs suffer from timing skew of at most ± 20 ps, then the region of concern is 20 ps. If there exists more than one maximum in this region, then there is no guarantee that the algorithm will converge to the right value of τ. A sufficient condition to ensure this is convexity of the correlation function within this region.

For application specific ADCs in which the signal autocorrelation function is known, the region of concern can be determined. However, generic bounds are useful, and can be derived by relating the autocorrelation to the signal power spectral density [28]. Assuming a differentiable autocorrelation and a real power spectral density,

$$R(\tau) = \int_{-\infty}^{\infty} G(f)e^{j(2\pi f)\tau} df \qquad (3.29)$$

Taking the derivative results in

$$\frac{dR(\tau)}{d\tau} = 2\pi j \int_{-\infty}^{\infty} f G(f)e^{j(2\pi f)\tau} df = -4\pi \int_{0}^{\infty} f G(f) \sin(2\pi f \tau) df \quad (3.30)$$

The region of concern is derived by having

$$\frac{dR(|\tau|)}{\tau} \leq 0 \qquad (3.31)$$

for all τ in $-\tau_{max} \leq \tau \leq \tau_{max}$. Thus, τ_{max} defines the region that guarantees a single maximum.

For example, if $G(f)$ is bandlimited to B such that $G(f) = 0$ for $f > B$, then $\tau_B = \frac{1}{2B} < \tau_{max}$, as seen from (3.30). In other words, if the expected timing skew is ± 20 ps, then an input signal bandlimited to 25 GHz is guaranteed to have a single maximum in this region. Although sufficient, such a condition is not necessary. An input signal with a first-order low pass power spectral density is not bandlimited, but has an autocorrelation function that is monotonically decreasing for $\tau > 0$, and thus has a single maximum for all τ.

3.5.3   Effect of Quantization

A final note in this section is on the effect of quantization, which was ignored in all the preceding analysis. The correlation was calculated through an approximation, and in (3.3) the variance of $E[M]$ only falls off with $1/M$ when it is uncorrelated [24]. This is not the case with quantization noise.

As a trivial example to illustrate this, assume an input signal of $x(t) = \cos(2\pi f_{in}t)$, where $f_{in} = f_s$. The output of the calibration ADC is $y_c[n] = \text{sign}(\cos(2\pi n)) = 1$ for all n. If the sub-ADC has single bit resolution, then its output is $y[n] = \text{sign}(\cos(2\pi(n - f_{in}\tau)))$, which is equivalent to

$$y[n] = \begin{cases} 1 & \text{if } \frac{1}{4f_{in}} \geq \tau \geq -\frac{1}{4f_{in}} \\ 0 & \text{else} \end{cases} \tag{3.32}$$

Thus, the value of the correlation does not change as long as $\frac{1}{4f_{in}} \geq \tau \geq -\frac{1}{4f_{in}}$, which means that the timing skew cannot be corrected because of the sub-ADC quantization. Note that this is not the case if the sub-ADC has infinite resolution such that $y[n] = \cos(2\pi(n - f_{in}\tau))$.

Therefore, quantization can be problematic. In stochastic signals, if the input signal $x(t)$ is stationary, ergodic, continuous, and has a non-zero probability of crossing $x(t) = 0$, then there is a non-zero probability that a zero-crossing exists between $nT_s - \tau$ and nT_s for all τ [29]. In other words, if enough samples are collected, then the number of zero-crossings decreases with τ, which is enough to ensure that the correlation function will increase and not suffer the effects of quantization.

This is not guaranteed in sinusoidal signals, as already illustrated. If the ratio of the input frequency to the sampling frequency is irrational, then collecting enough samples will guarantee that zero-crossings exists between $nT_s - \tau$ and nT_s for all τ, since there will exist samples where the value at nT_s and $nT_s - \tau$ do not have the same sign. If the ratio is rational, such that $\frac{f_{in}}{f_s} = \frac{N}{M}$ for integers N and M where the greatest common denominator of N and M is 1, then this is not guaranteed since the samples are periodic in M. However, if $\frac{1}{M} \leq \Delta\tau$, where $\Delta\tau$ is the delay line step size, then although the zero-crossings may not monotonically decrease as τ is continuously adjusted, they will decrease as τ is discretely adjusted with the delay line step size. As the resolution of the sub-ADC increases, this becomes less of a problem, and the number of frequencies in which the calibration algorithm will not properly work decreases.

3.6 Summary

In this chapter, sources of timing skew were discussed, and it was shown that the resulting timing skew is detrimental for high-speed input signals. A statistics-based background calibration algorithm was presented with analysis on the various aspects of the algorithm. The chapter concluded with some of the requirements on the input signal such that the algorithm functions properly.

Chapter 4
Architecture Optimization

A prototype ADC has been implemented as a proof-of-concept for the calibration algorithm presented in Chap. 3. An important phase in the design of ADCs is the high-level optimization, which allows design specifications to be met while either minimizing or maximizing an objective. For example, in flash ADCs, a common approach is to minimize the power dissipation of the comparator for a given sample rate while still meeting specifications on metastability rates, input-referred offset, input-referred thermal noise, kickback noise, and input capacitance.

Since a time-interleaved ADC, as discussed in Chap. 2, is used to achieve the high data rates required by serial links, the interleaving factor is an additional design parameter. It affects multiple parameters, such as sub-ADC sample rates, total area, total input capacitance, power, and design complexity, and results in a larger design space due to this extra degree of freedom.

This chapter presents a first-order optimization framework for time-interleaved flash ADCs and briefly extends it to real circuits. The results obtained for flash ADCs suggest an optimal interleaving factor, such that, given the technology parameters used, each flash ADC should operate in the low GS/s range.

4.1 Power Dissipation

Due to the low resolution requirements of serial links, each sub-ADC in the target design is a flash ADC. Excluding the track-and-hold and encoder circuitry, the main components of a flash ADC are the bank of comparators and the resistor ladder. Since serial links have power bounds, the objective of the optimization problem is to minimize the total power dissipation of the time-interleaved ADC. Ignoring second-order effects on power dissipation such as those resulting from clock distribution, the interleaving factor N directly relates the sub-ADC power $P_{\text{sub-ADC}}$ to the total time-interleaved ADC power P_{Total} such that

$$P_{\text{Total}} = N \cdot P_{\text{sub-ADC}} \tag{4.1}$$

M. El-Chammas and B. Murmann, *Background Calibration of Time-Interleaved Data Converters*, Analog Circuits and Signal Processing, DOI 10.1007/978-1-4614-1511-4_4,
© Springer Science+Business Media, LLC 2012

The power of each sub-ADC is a function of the number of comparators M, the comparator power P_{comp}, and the resistor ladder power P_{ladder}, as in

$$P_{sub-ADC} = M \cdot P_{comp} + P_{ladder} \qquad (4.2)$$

In a given flash ADC, M is a function of the ADC resolution B such that $M = 2^B - 1$, unless alternate architectures such as folding or subranging flash sub-ADCs are used [30, 31]. Furthermore, in the following analysis, the sampling period of the time-interleaved ADC is T_s, whereas the sampling period of each sub-ADC is $\hat{T}_s = N \cdot T_s$.

4.1.1 Dynamic Comparator First-Order Model

One way to minimize power is to use dynamic comparators, which not only have better sensitivity than CML latches [32], but are also more power efficient [33]. Dynamic comparators mainly dissipate power during the regeneration and reset phases, each of which lasts less than half the sub-ADC sampling period, $\hat{T}_s$. The following analysis does not include power due to reset for simplicity.

Assuming the comparator can regenerate within its allotted time of $\hat{T}_s/2$, the regeneration time is denoted by $\hat{T}_r$. The comparator power can be written as

$$P_{comp} = \frac{E_{comp}}{\hat{T}_s} \qquad (4.3)$$

where E_{comp} is the comparator energy. The comparator only conducts current during $\hat{T}_r$, such that $E_{comp} = E_r$, the energy consumed during regeneration. Substituting (4.2) and (4.3) in (4.1) results in

$$
\begin{aligned}
P_{Total} &= N \cdot P_{sub-ADC} \\
&= N \cdot \left(M \cdot P_{comp} + P_{ladder} \right) \\
&= N \cdot \left(M \cdot \frac{E_r}{\hat{T}_s} + P_{ladder} \right) \\
&= M \cdot f_s \cdot E_r + N \cdot P_{ladder} \qquad (4.4)
\end{aligned}
$$

since $f_s = N\hat{f}_s = \frac{N}{\hat{T}_s}$. Thus, the power due to the sub-ADCs is $M \cdot f_s \cdot E_r$, which only depends on the comparator energy, as both M and f_s are fixed.

An intuitive feel as to how the energy efficiency of a dynamic comparator changes can be obtained with a simple first-order model. A model similar to that in [34] is shown in Fig. 4.1, where the cross-coupled inverters can be linearized into a G_m circuit. Switches required for the setup and configuration of such a comparator are ignored, and the only capacitances, C_L, are those on the output nodes. This model is completely symmetric, no mismatches are included, and the latch is already placed in a region of instability before regeneration. The linearized inverters conduct current as long as their output nodes are not fully saturated to V_{DD} and ground. A

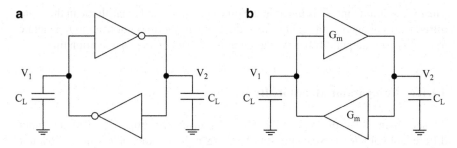

Fig. 4.1 (**a**) Cross-coupled inverter based dynamic latch. (**b**) Linearized cross-coupled inverter based dynamic latch

detailed derivation of the equations used in the remainder of the chapter is provided in Appendix B.

In Fig. 4.1, a differential voltage is applied to the nodes $V_1(t)$ and $V_2(t)$, such that

$$V_1(0) = V_c + v_d/2$$
$$V_2(0) = V_c - v_d/2 \tag{4.5}$$

where V_c is the common-mode voltage of the output nodes, v_d is the differential input signal, and $t = 0$ is the start of regeneration. Without loss of generality, it is assumed that $v_d > 0$.

The node voltages $V_1(t)$ and $V_2(t)$ for $t \geq 0$, as derived in Appendix B and shown in (B.18), are

$$0 \leq V_1(t) = \frac{v_d}{2}e^{(t/\tau)} + \frac{V_{DD}}{2} \leq V_{DD}$$

$$0 \leq V_2(t) = -\frac{v_d}{2}e^{(t/\tau)} + \frac{V_{DD}}{2} \leq V_{DD} \tag{4.6}$$

where $\tau = \frac{C_L}{G_m}$ is the regeneration time constant. The differential output voltage is

$$V_{od}(t) = V_1(t) - V_2(t) \tag{4.7}$$

such that

$$-V_{DD} \leq V_{od}(t) = v_d \cdot e^{(t/\tau)} \leq V_{DD} \tag{4.8}$$

Once the comparator is strobed, the input differential voltage v_d grows exponentially with a rate set by τ, until it is saturated to V_{DD}.

4.1.1.1 Dynamic Comparator Regeneration Time

The point at which the comparator completely regenerates to V_{DD} is derived with (4.8), such that

$$\hat{T}_r = \tau \ln\left(\frac{V_{DD}}{v_d}\right) \tag{4.9}$$

The regeneration time $\hat{T}_r$ is linear in the time constant τ and logarithmic in the input differential voltage v_d. In reality, since the comparator does not need to regenerate to a full-swing output, (4.9) serves as an upper bound on the regeneration time.

4.1.1.2   Comparator Metastability

As shown in (4.9), the regeneration time is inversely proportional to the input differential voltage v_d. Since the sub-ADC has a sample rate of $\hat{T}_s$, the comparator is said to be metastable [35] if $\hat{T}_r > \hat{T}_s/2$ since it would not have completely regenerated within its allotted time. The minimum acceptable input voltage is $v_{d,m}$ such that

$$v_{d,m} = V_{\mathrm{DD}} \cdot e^{-\hat{T}_s/(2\tau)} \tag{4.10}$$

and thus the metastability rate, or probability that a comparator is metastable, assuming a uniform input signal distribution and a full-scale input signal of V_{DD}, is

$$MR = P(\text{comparator is metastable}) = \frac{v_{d,m}}{V_{\mathrm{DD}}} = e^{-\hat{T}_s/(2\tau)} \tag{4.11}$$

which is inversely proportional to the sampling period.

4.1.2   Dynamic Comparator Power

The power dissipated in the dynamic comparator results from the total current drawn from the power supply. This current equals the sum of the currents the PMOS transistors in each linearized inverter conduct, which is derived in Appendix B to be

$$I_{V_{\mathrm{DD}}}(t) = \begin{cases} \frac{G_m}{2} \cdot V_{\mathrm{DD}} & \text{if } 0 \leq t \leq \hat{T}_r \\ 0 & \text{else} \end{cases} \tag{4.12}$$

The power dissipated is

$$P_{\mathrm{comp}} = \frac{V_{\mathrm{DD}}^2}{\hat{T}_s} \cdot \left(\frac{C_L}{2} \cdot \ln\left(\frac{V_{\mathrm{DD}}}{v_d} \right) \right) \tag{4.13}$$

Therefore, the comparator energy, as in (4.4), is

$$E_r = P_{\mathrm{comp}} \cdot \hat{T}_s$$

$$= V_{\mathrm{DD}}^2 \cdot \left(\frac{C_L}{2} \cdot \ln\left(\frac{V_{\mathrm{DD}}}{v_d} \right) \right) \tag{4.14}$$

which is directly proportional to C_L and V_{DD} and inversely proportional to the input differential voltage v_d. Equation (4.4) becomes

$$P_{\text{Total}} = (M \cdot f_s) \cdot \left(\frac{C_L}{2} \cdot V_{DD}^2 \cdot \ln \left(\frac{V_{DD}}{v_d} \right) \right) + N \cdot P_{\text{ladder}} \qquad (4.15)$$

The design parameters that affect the total power dissipation are the load capacitance, the interleaving factor, and the power in the resistor ladder. The other terms in (4.15), such as M, f_s, V_{DD}, and v_d tend to be fixed for a given design.

4.2 First-Order Optimization Framework

The circuit parameters that affect the total time-interleaved ADC power, as in (4.15) are M, the number of comparators in each flash sub-ADC, C_L, the load capacitance, and P_{ladder}, as set by the resistor ladder impedance. This section completes the optimization setup and develops a set of constraints such that the minimum power is realizable.

An assumption used in the derivation of the sub-ADC power was that the comparator had completely regenerated, resulting in a constraint on $\hat{T}_r$, as derived in (4.9). Since each sub-ADC has a period of $\hat{T}_s$, such that there are $N = \frac{T_s}{\hat{T}_s}$ interleaved sub-ADCs, $\hat{T}_r \leq \hat{T}_s/2$ is a necessary constraint, assuming the sub-ADC clock has a 50% duty cycle.

$\hat{T}_r$ is linear in τ, as in (4.9), which is a function of C_L and G_m, both of which can be divided into several factors. To a first-order, the G_m of the linearized inverter linearly increases with width, such that $G_m = G_{m,0} W_{\text{inv}}$, where $G_{m,0}$ is the transconductance for a width of $1\,\mu\text{m}$ and W_{inv} is the width of the inverter. The load capacitance can be divided into the inverter's intrinsic capacitance C_I, due to the transistors within the inverter, and the extrinsic capacitance C_E, due to various loads and traces. Therefore, $C_L = C_I + C_E$, where C_E can be assumed to be fixed. The intrinsic capacitance is $C_I = C_{I,0} W_{\text{inv}}$, since it also increases linearly with the width of the inverter. Thus, (4.9) is rewritten as

$$\hat{T}_r = \left(\frac{C_{I,0} W_{\text{inv}} + C_E}{G_{m,0} W_{\text{inv}}} \right) \cdot \ln \left(\frac{V_{DD}}{v_d} \right) \qquad (4.16)$$

and (4.15) becomes

$$P_{\text{Total}} = (M \cdot f_s) \cdot \left(\frac{(C_{I,0} W_{\text{inv}} + C_E)}{2} \cdot V_{DD}^2 \cdot \ln \left(\frac{V_{DD}}{v_d} \right) \right) + N \cdot P_{\text{ladder}} \qquad (4.17)$$

both of which are a function of the inverter width W_{inv}. In a first pass of the analysis, P_{ladder} is set to zero.

4.2.1 Performance Limits

Two limits can be derived from (4.16) as a function of W_{inv}, the width of the inverters. The first limit is derived when W_{inv} tends to 0, which corresponds to extremely small cross-coupled inverters, and results in $C_{I,0} W_{inv} \ll C_E$. Therefore,

$$\hat{T}_r \approx \left(\frac{C_E}{G_{m,0} W_{inv}} \right) \cdot \ln \left(\frac{V_{DD}}{v_d} \right) \tag{4.18}$$

$$P_{Total} \approx (M \cdot f_s) \cdot \left(\frac{C_E}{2} \cdot V_{DD}^2 \cdot \ln \left(\frac{V_{DD}}{v_d} \right) \right) \tag{4.19}$$

$\hat{T}_r$ is inversely proportional to the inverter width, whereas the total power is constant and presents a lower bound on the minimum power dissipation. Thus, sizing down the comparator increases its regeneration time and leads to diminishing returns in power savings. In the second limit, the inverter width is increased such that $C_{I,0} W_{inv} \gg C_E$ and

$$\hat{T}_r \approx \frac{C_{I,0}}{G_{m,0}} \ln \left(\frac{V_{DD}}{v_d} \right) \tag{4.20}$$

$$P_{Total} \approx (M \cdot f_s) \cdot \left(\frac{C_{I,0}}{2} \cdot W_{inv} \cdot V_{DD}^2 \cdot \ln \left(\frac{V_{DD}}{v_d} \right) \right) \tag{4.21}$$

In this case, the regeneration time is constant, whereas the power required is directly proportional to width. Thus, regardless of how much power is consumed by the dynamic comparator, a technological wall is reached that prevents the comparator from regenerating faster.

4.2.2 Optimization Analysis

The objective function and the constraint on the regeneration time can be combined into

$$\min_{W_{inv}} \quad P_{Total}$$

$$s.t. \quad \hat{T}_r \leq \frac{\hat{T}_s}{2} \tag{4.22}$$

The objective function in (4.22) consists only of the power of the comparators, since $P_{ladder} = 0$ at this stage of the analysis. Thus,

$$\min_{W_{inv}} \quad (M f_s) \left(\frac{(C_{I,0} W_{inv} + C_E)}{2} \cdot V_{DD}^2 \cdot \ln \left(\frac{V_{DD}}{v_d} \right) \right)$$

$$s.t. \quad \left(\frac{C_{I,0} W_{inv} + C_E}{G_{m,0} W_{inv}} \right) \cdot \ln \left(\frac{V_{DD}}{v_d} \right) \leq \frac{\hat{T}_s}{2} \tag{4.23}$$

Due to this being a convex optimization problem in W_{inv}, the constraint will hold with equality [23], and the optimal objective function becomes

$$P_{\text{Total}} = \left(\frac{M f_s}{2}\right) \cdot V_{\text{DD}}^2 \cdot \left(C_E + 2 C_{I,0} \left(\frac{C_E \ln\left(\frac{V_{\text{DD}}}{v_d}\right)}{\hat{T}_s G_{m,0} - 2 C_{I,0} \ln\left(\frac{V_{\text{DD}}}{v_d}\right)} \right) \right) \cdot \ln\left(\frac{V_{\text{DD}}}{v_d}\right)$$

(4.24)

Equation (4.24) is a strictly decreasing function in the sub-ADC sampling period $\hat{T}_s$. Increasing the sampling period, and thus increasing the interleaving factor, reduces the overall power consumption of the time-interleaved ADC. This converges to the first performance limit in (4.19).

4.2.2.1 Example

To illustrate the relationship between the time-interleaved ADC power and the sub-ADC sampling period, we set the technology parameters of the first-order comparator to be $f_T = 300\,\text{GHz}$ and $G_{m,0} = 300\,\mu\text{S}/\mu\text{m}$ such that $C_{I,0} = 1\,\text{fF}/\mu\text{m}$. The design specifications are a sample rate of $f_s = 10\,\text{GS/s}$, a power supply voltage of $V_{\text{DD}} = 1\,\text{V}$, and a metastability rate of $MR = 10^{-9}$. The external capacitance on the voltage nodes of the comparators is $C_E = 5\,\text{fF}$. Furthermore, if a 5 bit ADC is used, then $M = 2^B - 1 = 31$ comparators.

With these values, it is possible to plot the optimal inverter width and power as a function of N, the interleaving factor, as in Fig. 4.2a,b. The area above the curve in both plots is the feasible region. As is expected, the optimal width and optimal power dissipation monotonically decrease as the interleaving factor increases. Furthermore, as a result of the parameter values chosen, for the given metastability rate, at least two sub-ADCs are required. This is shown in Fig. 4.3a, which plots the minimum acceptable interleaving factor as a function of power such that the metastability rate is met. With a higher metastability rate, a single channel is possible, as shown in Fig. 4.3b, which uses a metastability rate of 10^{-6}. Figure 4.4 plots the relationship between the metastability rate and the minimum acceptable interleaving factor.

Even though the power savings increase with the interleaving factor, as in Fig. 4.2b, these savings become marginal and result in diminishing returns, especially when the design complexity is considered. Interleaving several hundred sub-ADCs is possible [36], but the improvement in power is not necessarily worth the overhead in both complexity and area.

4.2.2.2 Example with Resistor Ladder

When the power of the resistor ladder is included such that $P_{\text{ladder}} \neq 0$, there is a clearly optimal interleaving factor. The resistor ladder dissipates static power and is set by the impedance of the resistor ladder. Furthermore, the total power consumed

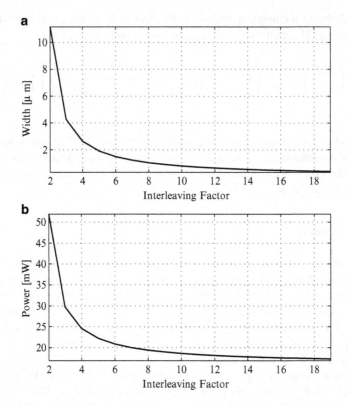

Fig. 4.2 (**a**) Optimal width for the first-order comparator model. (**b**) Optimal time-interleaved ADC power

by all the resistor ladders in the time-interleaved ADC is directly proportional to N, whereas the sub-ADC power is inversely proportional to N. These two competing factors result in an optimal interleaving factor that minimizes the total power. Figure 4.5 plots the total power for different values of the resistor ladder, and there is a minimum in all three cases.

4.2.2.3 Framework Limitations

In a more realistic optimization framework, other constraints, such as those on the minimum or maximum possible widths, comparator offset, input-referred noise, and clock distribution power, are included. Just like the resistor ladder, these would prevent the line in Fig. 4.2a from strictly decreasing.

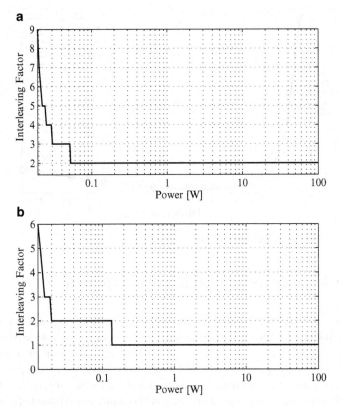

Fig. 4.3 Smallest possible interleaving factor for a given power dissipation with a metastability rate of (**a**) 10^{-9} and (**b**) 10^{-6}

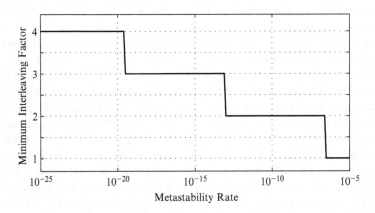

Fig. 4.4 Smallest possible interleaving factor as a function of metastability

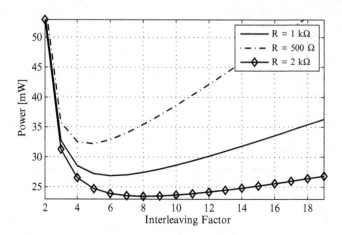

Fig. 4.5 Optimal power with resistor ladder

4.3   A Circuit-Oriented Optimization Approach

The first-order model presented in the previous section provides an intuitive understanding of the relationship between the comparator sizing, interleaving factor, and power. Deriving analytic equations describing the operation of a transistor-based comparator, as opposed to the first-order model used in this chapter, with an accuracy comparable to simulation is nontrivial. An alternate approach is to make use of CPU power and to design a circuit-based optimization framework. With the current availability of computational power, this approach is attractive, although knowledge of the underlying circuitry is necessary to keep the problem tractable. Furthermore, this also enables the designer to compare various architectures, which will require different analytic equations, and to include manufacturing variations in the simulations.

The procedure is to parameterize the different components of the comparator that the designer cares about. This can include transistor widths, as well as setup voltages such as the input common-mode. A brute-force approach is possible, in which all possible permutations of parametric values are used, but this becomes exponentially unwieldy in terms of computation time. For example, if there are five variables to optimize with ten possible points each, a total of 100,000 simulations are needed. Other less computationally expensive methods assume some form of convexity in the optimization problem, which in many cases is a reasonable assumption. For example, Fig. 4.6 plots the power of a time-interleaved ADC as a function of the interleaving factor using simulation data for a single comparator. The variables in the comparator circuit are the widths of the various transistors in Fig. 5.6, and are explained further in Appendix C. Excluding the power of the resistor ladder, the optimal curve of Fig. 4.6 resembles that of Fig. 4.2b, and clearly decreases as a function of the interleaving factor.

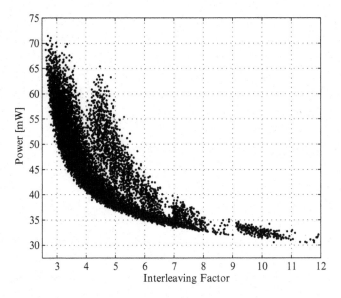

Fig. 4.6 Simulated time-interleaved ADC power with different comparator sizings

4.4 Summary

This chapter presented a framework for the optimization of time-interleaved ADCs. The results show that for interleaved flash ADCs, there is an optimal value for the interleaving factor, which is a function of the load capacitance of the dynamic comparators, the ADC resolution, the sampling rate, and the static power dissipation of the resistor ladder. An extension to transistor-level circuits was discussed, and the plotted results have a similar form to those of the high-level framework.

Summary

Chapter 5
Circuit Design

The architecture of the prototype ADC designed to evaluate the calibration algorithm of Chap. 3 is presented in Fig. 5.1. The main components of this ADC are the array of interleaved sub-ADCs, the eight delay lines, the calibration ADC required by the background calibration algorithm, the phase generator, and an off-chip digital calibration block. Additional circuitry required to interface the prototype with test equipment include the low-voltage differential signaling (LVDS) output drivers. This chapter details the design of these blocks.

5.1 The Sub-ADC

Each sub-ADC is a 5 bit flash ADC. Based on the architecture optimization procedure discussed in Chap. 4, an interleaving factor of eight is chosen. Thus, the sub-ADCs have a sample rate of 1.5 GS/s each, resulting in an aggregate sample rate of 12 GS/s for the time-interleaved array. As shown in Fig. 5.2, the sub-ADC consists of a bootstrapped track-and-hold, a bank of comparators, a resistor ladder, and a Wallace Encoder.

5.1.1 Bootstrapped Track-and-Hold

Due to the inherent sampling nature of dynamic comparators, track-and-holds are not necessarily required in flash ADCs [37], as long all the comparators in the ADC sample the input signal at the same time. However, variations in the sampling times of each comparator exist and result in sampling errors that grow with input signal frequency, as derived in Appendix D. Since a high frequency input signal is expected, a track-and-hold is used to remove the effects of comparator skew.

M. El-Chammas and B. Murmann, *Background Calibration of Time-Interleaved Data Converters*, Analog Circuits and Signal Processing, DOI 10.1007/978-1-4614-1511-4_5, © Springer Science+Business Media, LLC 2012

Fig. 5.1 Prototype ADC
architecture

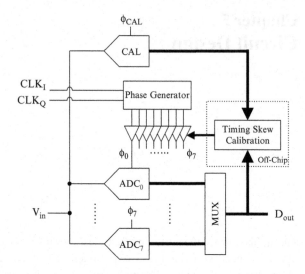

Fig. 5.2 Sub-ADC block
diagram

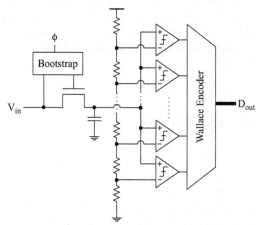

The track-and-hold subsamples the input signal and requires an acquisition bandwidth commensurate with the input signal bandwidth. However, as the sub-ADC resolution is 5 bits, an active track-and-hold is not necessary, simplifying its design and reducing its power consumption. Although a single NMOS switch followed by a sampling capacitor is an attractive candidate for a passive track-and-hold, it does not provide sufficient linearity at high frequencies. Furthermore, its performance is also dependent on the input common-mode and on the input signal amplitude [38]. Figure 5.3 plots the change in the simulated output signal-to-distortion ratio (SDR) of the NMOS switch as a function of the input signal amplitude for a 6 GHz signal, optimized for input common-mode voltage and sampling capacitance. As is expected, the SDR decreases with increasing amplitude [38], and barely reaches the 5 bit performance with 0.3 V input amplitude.

To first-order, bootstrapping the switch [39] separates the linearity performance of the track-and-hold from the input common-mode and amplitude due to the

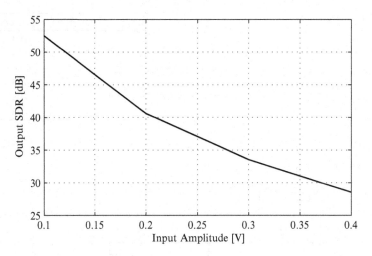

Fig. 5.3 Output SDR results of NMOS sampling switch with a 6 GHz input signal

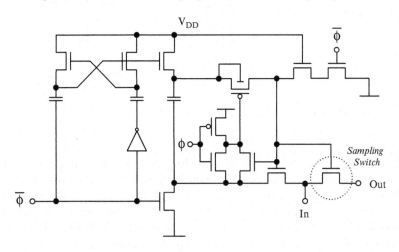

Fig. 5.4 Track-and-hold schematic

improved resistor linearity, and is a more practical solution with high frequency input signals. The bootstrapped circuit is implemented as in Fig. 5.4 [40], which results in a low-power and reliable passive track-and-hold.

A general concern in sampling circuits is thermal noise, which has a variance of kT/C [11, 41, 42]. A minimum acceptable capacitance $C_{\min}$ can be derived by setting the thermal noise variance to be less than a quarter of the quantization noise variance, such that

$$C_{\min} = \frac{48 \cdot kT \cdot 2^{2B}}{A^2} \qquad (5.1)$$

Table 5.1 Capacitance sizing

ADC resolution [Bits]	Capacitor [fF]
3	0.035
4	0.14
5	0.56
6	2.25
7	8.98
8	35.93

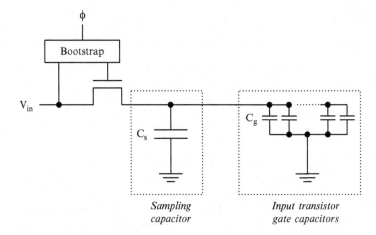

Fig. 5.5 Track-and-hold with sampling capacitances

Table 5.1 displays the values of the sampling capacitor as a function of ADC resolution for an input peak-to-peak voltage of $A = 0.6$ V and room temperature of 25°C. A 5 bit ADC requires a capacitance of less than 1 fF. Given that the track-and-hold is loaded by the input capacitances of 31 comparators and that thin-oxide devices in a 65 nm process have a gate capacitance of approximately 1 fF/μm, kT/C noise does not set the size of the sampling capacitor.

Clock-feedthrough and charge injection can be an issue as they degrade the sampled signal. Adding an explicit sampling capacitor to the track-and-hold output node, as in Fig. 5.5, where C_s is the extra sampling capacitor and the parallel capacitors C_g are those due to the gate capacitance of the comparators, helps reduce these problems. The extra capacitor C_s has an upper bound due to the increased time constant of the sampling network. Sizing up the sampling switch is an option but this increases the resulting clock-feedthrough and also increases the power dissipation of the track-and-hold. In this design, $C_s = 110$ fF and $C_g = 3$ fF.

Fig. 5.6 Schematic of dynamic comparator

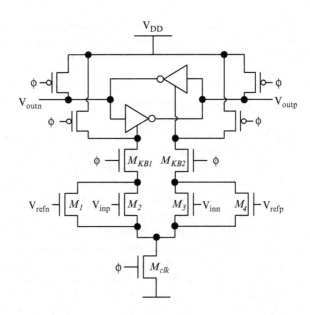

5.1.2 Comparator Design

Following the track-and-hold described in Sect. 5.1.1 is a bank of 31 comparators. Dynamic comparators are used as they have higher input sensitivity, higher energy efficiency, and a smaller latching time constant than CML latches, thus increasing their attractiveness when designing for high-speed input signals [32, 33].

The dynamic comparator is sense-amplifier based [43], as in Fig. 5.6. The complete implementation includes a chain of inverters at the output of the comparator to asynchronously increase the latch gain, thus decreasing the metastability rate [35, 44], and to separate the input load of the next stage from the comparator output nodes, as this directly affects the comparator speed.

The transistors in the dynamic comparator can be divided into three groups. The first is the pair of cross-coupled inverters, which results in a positive feedback loop. The regeneration time constant is a function of the output load capacitance and the transconductance of the cross-coupled pair, as explained in Chap. 4. The second group is the series of parallel transistors M_{1-4}, which connect the comparator to the differential input signal and the comparator reference voltages. The comparator output swings in the direction set by these voltages, such that

$$V_{\text{out,diff}} = V_{\text{DD}} \cdot \text{sign}\left[\left(V_{\text{inp}} - V_{\text{inn}} \right) - \left(V_{\text{refn}} - V_{\text{refp}} \right) \right] \tag{5.2}$$

The third group consists of three clocked transistors, M_{clk}, M_{KB1}, and M_{KB2}, as well as the PMOS reset transistors. When the clock ϕ goes high, these first three transistors turn on and conduct current, allowing the comparator to regenerate. M_{clk}

also offers a degree of freedom when designing for input-referred offset, while M_{KB1} and M_{KB2} are inserted to reduce kickback. When the clock goes low, the reset transistors turn on and pull the nodes up to V_{DD}.

5.1.2.1 Design Considerations

In addition to metastability, which entails sizing the transistors to increase the overall comparator gain, three additional design considerations are input-referred noise, input-referred offset, and kickback noise.

Although low-resolution ADCs do not generally suffer from thermal noise issues due to their large quantization error, noise should still be investigated to ensure that it does not result in SNDR degradation. Nuzzo et al. [45] present noise analysis based on stochastic differential equations for a comparator similar to that in Fig. 5.6, and concludes with several rule of thumb design techniques to reduce noise.

In the comparator optimization of Chap. 4, input-referred noise was added as a design constraint in the optimization framework. This was included in the optimization framework as discussed in Chap. 4 with the use of the SpectreRF™ PNOISE analysis [46]. In the resulting design space, input-referred noise was not a limiting factor.

On the other hand, input-referred offset was a severe limiting factor, as it presented a lower bound on the comparator power dissipation. Offset is due to both static mismatch such as threshold variations as defined by the Pelgrom model [15] and dynamic mismatch, such as capacitive variations [34]. Analytic relationships of the transistor variations and the input-referred offset for a given comparator architecture are possible [47], but lose accuracy as they ignore various circuit parameters such as input common-mode voltages.

In sizing the transistors to reduce offset, design guidelines from Nuzzo et al. [45] can be used by modeling input-referred offset as low frequency input-referred noise [48]. The sources of offset in the comparator can be divided into two groups. The first is that due to the input pair, and their offset can be reduced by increasing their size. However, this also increases the input capacitance and dynamic power consumption. The second group of sources of offset is due to the kickback and inverter transistors. As shown in [45], the input-referred offset due to this group is directly proportional to the overdrive voltage of the input pair and to the discharge current. Thus, given the required full-scale range of the input signal, the minimum acceptable overdrive voltage is used, and the size of the clock transistor M_{clk} is reduced such that the discharge current decreases. Input-referred offset is added to the optimization framework of Chap. 4 by calculating the offset via simulations, as in [49].

The final design consideration is kickback noise, which results in disturbances on the input and reference nodes due to swings on the drain nodes of the transistors M_{1-4} [50]. Inserting the pair of clocked transistors M_{KB1} and M_{KB2} between the input and reference transistors and the cross-coupled inverter, as in Fig. 5.6, reduces kickback by preventing the precharge of the drain nodes [51].

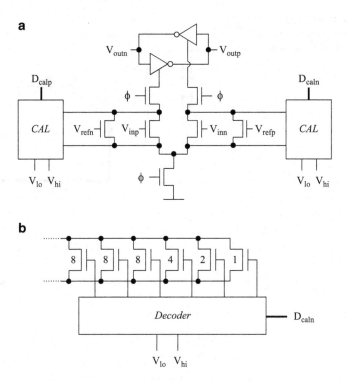

Fig. 5.7 (**a**) Dynamic comparator with offset correction. Reset transistors are not shown. (**b**) Segmented calibration DAC with relative transistor widths

5.1.2.2 Comparator Offset Correction

Decreasing the widths of the comparator transistors reduces power dissipation because of the smaller node capacitances. A side benefit of this is a reduction of the input capacitance. However, it also leads to increased threshold variation [15]. Although it is possible to size the comparator such that performance yield constraints are met, such an approach is power-inefficient. An alternate approach is to provide the comparator with a trim DAC that compensates for input-referred offset in order to meet yield constraints [52–54].

As shown in Fig. 5.7a, an approach similar to [54] is used in this work. A 5 bit calibration DAC is placed parallel to the input and reference transistors such that it compensates for the comparator offset by differentially injecting current in the two comparator branches. By varying the differential current, the calibration DAC biases the comparator in a direction that overcomes the effect of offset. The calibration DAC consists of parallel transistors, as in Fig. 5.7b, and is segmented with three binary encoded bits and two thermometer encoded bits to improve monotonicity [55].

Fig. 5.8 (a) Foreground
offset correction. (b) Timing
diagram for foreground offset
correction

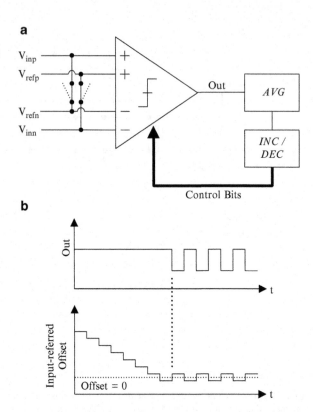

Due to the parallel placement of the calibration DAC to the input and reference
transistors, the large LSB size, and the ratiometric behavior of the comparator
[45, 56], the required calibration code that compensates for the offset is to first-
order temperature independent, ignoring dynamic effects. This allows the use of a
foreground offset calibration technique that is run at the system startup.

As in [53], the calibration code required by the DAC in Fig. 5.7b is incremented
by a single LSB with every update. A calibration engine is designed on-chip for
each sub-ADC, and includes the option of controlling the calibration off-chip.
Each comparator is calibrated by shorting the input and reference transistors, as
in Fig. 5.8a. In the absence of offset, this biases the comparator at its switching
point. The output is averaged over multiple cycles to remove the errors due to
thermal noise. In this work, the on-chip calibration engine averaged four cycles for
each update, and this number can be changed by running the calibration off-chip.
Depending on the output of the averaging block, the control bits for the calibration
DAC are either incremented or decremented, in a direction that decreases the input-
referred offset, which results in the timing diagram of Fig. 5.8b. As the control bits
are adjusted, the input-referred offset converges to zero. Once it crosses zero, the
output of the comparator changes values (e.g. from zero to one) and the adjustment

direction of the control bits flips. Thus, the input-referred offset hovers around zero while the comparator output oscillates between zero and one. The residual offset is a function of the calibration DAC resolution and of second-order mismatch effects.

5.1.3 Resistor Ladder

Although it is possible to intentionally imbalance the comparator such that it inherently creates different switching points [57], the degraded power supply sensitivity leads to increased input-referred supply noise. Thus, a resistor ladder is implemented that differentially creates the reference voltages for all 31 comparators. As has been discussed, kickback noise affects the reference levels of the resistor ladder. A power-inefficient solution is to decrease the impedance of the resistor ladder, as the RC time constant of the reference nodes decreases. This allows each node to settle fast enough as to not disturb the next sample. In order to avoid this power penalty, and as discussed in Sect. 5.1.2, a pair of clocked transistors have been placed between the cross-coupled inverters and the input and reference transistors [51], as in Fig. 5.6.

5.1.4 Wallace Encoder

An encoder is used to represent the binary outputs of the 31 comparators with a 5 bit word. The prototype ADC implements a Wallace Encoder [58], which is a ones-adder that sums the outputs of the comparators and which has a lower error-rate than other commonly used encoders [59]. One drawback of this approach is the power consumption, which exponentially increases with the ADC resolution, and which is unwieldy in older technology nodes [60]. However, for a 5 bit ADC implemented in a 65 nm technology, the use of a Wallace Encoder is acceptable.

The Wallace Encoder follows a straightforward logical scheme that recursively implements a 3-2 encoder, 7-3 encoder, a 15-4 encoder, and a 31-5 encoder. The basic unit of this encoder is a full-adder, which takes as inputs three bits A, B, and C_i, which is the input carry bit, and outputs a sum bit S and a carry bit C_o, as in Table 5.2.

A $(2^N - 1)$-N encoder is recursively built by taking two $(2^{N-1} - 1)$-$(N - 1)$ encoders and independently adding their output sum bits and carry bits [61]. Thus, a 7-3 encoder is created by taking two 3-2 encoders (each of which is a full-adder) and combining them with two additional full-adders, as shown in Fig. 5.9.

This is then extended into a 15-4 encoder by again combining two 7-3 encoders, as in Fig. 5.10. And again, this is recursively extended to a 31-5 Wallace Encoder. In general, the number of full-adders required for a B bit ADC is $2^B - B - 1$ [62], which is responsible for the exponential power increase.

Table 5.2 Full-adder
operation

C_i	B	A	C_o	S
0	0	0	0	0
0	0	1	0	1
0	1	0	0	1
0	1	1	1	0
1	0	0	0	1
1	0	1	1	0
1	1	0	1	0
1	1	1	1	1

Fig. 5.9 7-3 Wallace
Encoder

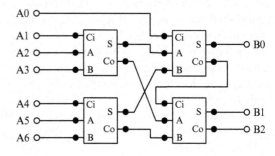

Fig. 5.10 15-4 Wallace
Encoder

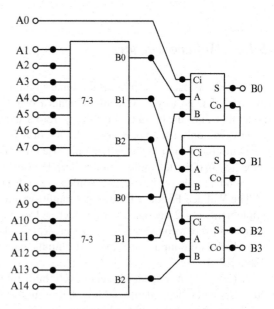

5.2 The Delay Line

As discussed in Chap. 3, the delay line provides an adjustable knob that enables
the calibration algorithm to minimize the timing difference between the clocks of
each sub-ADC and the calibration ADC, which in this implementation consists of a

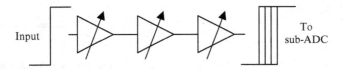

Fig. 5.11 Variable delay line consisting of cascaded delay cells

Fig. 5.12 Variable delay cell

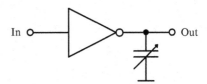

single comparator. It is designed to have a correction range that covers the expected delay variations such that yield constraints are met and a correction step size that reduces the timing skew to less than the design specifications, as derived in Chap. 2.

The 7-bit delay line used in this prototype consists of a series of cascaded delay cells, as in Fig. 5.11, and the resulting simulated range and step size were approximately 32 ps and 0.25 ps, respectively.

5.2.1 The Delay Cell

The basic block of each delay cell is an inverter, which has low dynamic power consumption. The delay of the inverter is adjusted with a variable capacitor, as in Fig. 5.12. Since these delay cells lie directly in the clock path, performance requirements, in addition to the calibration range and step size, include thermal [63] and power-supply noise jitter. Although inverter jitter performance can be improved through design [64, 65], inverters have poor supply rejection [66] as there is almost a one-to-one correspondence in the change of voltage supply to the change in delay. For example, a 10% change in supply will result in a 10% change in delay, which can easily result in several picoseconds of jitter, given an FO4 delay for a 65 nm process of approximately 25 ps. A common solution for this is to stabilize the inverter power supply with a voltage regulator. The approach used in this prototype was to provide the delay lines with separate power and ground lines from off-chip.

5.2.1.1 Variable Capacitive Load

The delay of an inverter is a function of both the inverter drive strength and its load capacitance. A first-order model of an inverter has a delay of

$$t_{\text{delay}} = C_L \frac{V_d}{I_{\text{inv}}} \tag{5.3}$$

Fig. 5.13 Delay cell with
capacitive load

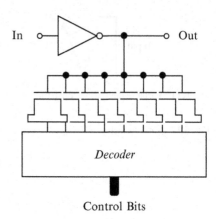

where C_L is the load capacitance, $V_d = V_{DD}/2$, and I_{inv} is the inverter current,
which results in a linear relationship between the load capacitance C_L and the delay
t_{delay}. Thus, changing C_L by ΔC_L changes the delay, such that

$$\Delta t_{delay} = \Delta C_L \frac{V_{DD}}{2I_{inv}} \qquad (5.4)$$

Δt_{delay} is a function of the inverter drive strength, and for a given change in
capacitance ΔC_L, the change in delay Δt_{delay} can be decreased by sizing up
the inverter. Thus, the inverter and capacitive load are codesigned by choosing
the appropriate inverter strength, which results in current I_{inv}, and the capacitive
change ΔC_L in order to achieve the required minimum step size, given technology
limitations and process variations.

This variable capacitor is built using the gate capacitance of MOS transistors
[21], which is a function of the transistor bias voltages. Shorting the drain and
source node of the transistor and digitally controlling this shorted node changes
the gate capacitance, and in turn, the delay of the inverter. In this work, a fully
controllable load is created with a 7-bit array of digitally controlled MOS transistors,
as in Fig. 5.13, and is segmented with five binary encoded bits and two thermometer
encoded bits. The minimum change in capacitance, ΔC_L, was approximately 0.6 fF.

5.2.2   Cascaded Delay Cells

The delay line is divided into several delay cells, as in Fig. 5.11, in order to
minimize effects of thermal and power-supply jitter by limiting the change in delay
to approximately 30% of the inverter delay. As shown in Fig. 5.14, each delay cell is

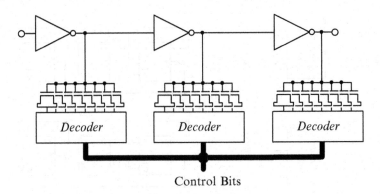

Fig. 5.14 Complete variable delay line

controlled by the same 7-bit control word, such that the delays of each cell always change in the same direction. This helps improve the delay line monotonicity.

5.3 Phase Generator

The clocks for each sub-ADC are created with a phase generator. Many designs use a PLL or DLL for this purpose [21]. However, it is also possible to use a shift register to create the sub-ADC clocks [67]. In this design, cascaded shift registers are used. Since the resulting sub-ADC clocks need to be spaced with a phase offset of 45°, it is helpful to bring in two signals with twice the required frequency and a 90° phase offset. These in phase and quadrature phase differential clocks pass through four series of cascaded shift registers, as in Fig. 5.15. Each group of shift registers has a divide-by-two block, such that the two outputs of each group both have the required frequency of 1.5 GHz. Thus, the eight outputs have the required timing offset for an 8-way time-interleaved ADC.

5.4 Output Buffers

The ADC outputs are transmitted off chip using LVDS. Although each output signal requires two pins, LVDS enables higher speed transmission than regular CMOS input/output cells, and also dissipates less power since low-voltage signals are used [68, 69]. LVDS output voltages are specified with a common-mode voltage of $1.125\,\text{V} \leq V_{CM} \leq 1.375\,\text{V}$ and a differential voltage of $0.25\,\text{V} \leq |V_{diff}| \leq 0.45\,\text{V}$. The LVDS output buffer used in the prototype ADC consists of a level converter and an LVDS driver.

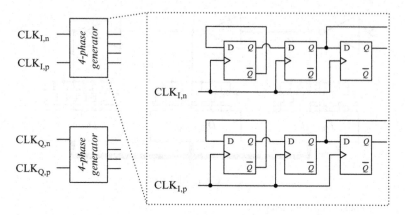

Fig. 5.15 Phase generator for sub-ADC clocks

Fig. 5.16 Level converter

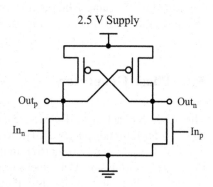

5.4.1   Level Converter

The nominal voltage for the prototype ADC is 1 V. However, the LVDS drivers run off a voltage supply of 2.5 V, and the architecture used requires input signal swings between 0 V and 2.5 V. The conversion from 1 V digital signals to 2.5 V signals is achieved with a latch-based structure, as in Fig. 5.16. This level converter takes as inputs a digital bit and its complement, such that one of the NMOS transistors is turned off. The remaining NMOS transistor conducts current and allows the cross-coupled PMOS transistors to regenerate such that the outputs are pulled to 2.5 V and 0 V. This converts the digital voltage levels for the next stage.

5.4.2   LVDS Driver

The LVDS driver consists of a transmitter and a closed-loop control circuit that keeps the output common-mode voltage within specifications. The transmitter [69]

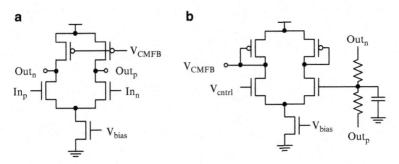

Fig. 5.17 (a) LVDS transmitter. (b) LVDS common-mode feedback control circuit

is shown in Fig. 5.17a. The bias voltage for the PMOS transistors is V_{CMFB} and comes from the closed-loop control circuitry in Fig. 5.17b, which resistor-averages the transmitter outputs and compares it to a reference voltage V_{cntrl}. This ensures the common-mode voltage is within the required bounds.

5.5 Summary

In this chapter, the design for the prototype ADC was detailed. The different components of each of the eight sub-ADCs, including foreground offset calibration for the comparators, and the design of the eight delay lines were discussed in detail. Furthermore, the phase generator, which creates the sub-ADC clocks, and the output buffers, were both presented.

Chapter 6
Measurement Results

The prototype ADC discussed in Chap. 5 was fabricated to evaluate the background calibration algorithm described in Chap. 3. This chapter presents the test setup and the ADC measurement results.

6.1 Test Setup

The test setup used to gather measurement data for the prototype ADC is shown in Fig. 6.1, and the test equipment models are listed in Table 6.1. The test setup consists of the device under test (DUT), the printed circuit board (PCB), several signal generators for the input signal, sub-ADC clocks, and the calibration ADC clock, two data capture cards, and a computer which runs the background timing skew calibration algorithm.

6.1.1 Device Under Test

The prototype ADC was implemented in TSMC 65 nm GP, and has a total and active area of 1.3 mm^2 and 0.44 mm^2, respectively. It has a total of 45 pins, and the die photo of this prototype is shown in Fig. 6.2. Each of the eight drawn rectangles outlines a single sub-ADC. The die was packaged in a QFN-48 package.

6.1.2 Printed Circuit Board

A four-layer PCB was used in order to include a ground and power distribution plane. The PCB provided an interface with the data capture cards, the signal

M. El-Chammas and B. Murmann, *Background Calibration of Time-Interleaved Data Converters*, Analog Circuits and Signal Processing, DOI 10.1007/978-1-4614-1511-4_6, © Springer Science+Business Media, LLC 2012

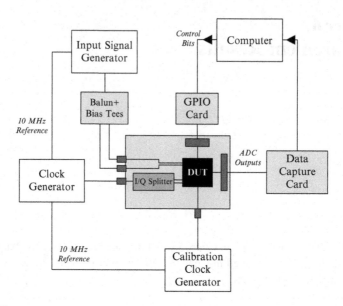

Fig. 6.1 Test setup

Table 6.1 Test equipment
used in Fig. 6.1

Use	Part number
Clock generator	HP 83711B
Input signal generator	HP 83732B
Calibration clock generator	HP 8664A
I/Q splitter	QCN-45+
Data capture card	TI TSW1200
GPIO card	NanoRiver miniboard

generators, and the voltage supplies. Since the DUT requires differential in phase
and quadrature phase clocks, as discussed in Chap. 5, a power divider was included
on the board to create two signals with an approximately 90° phase shift, each
followed by a transformer that created differential signals. An option was included
to bypass this approach, such that the differential in phase and quadrature phase
clocks could be brought from off-board.

Although a similar option could have been used for the input signal, using a single on-board transformer for the large range of input frequencies was problematic,
and an off-board balun and bias tee combination was used to create the differential
input signal.

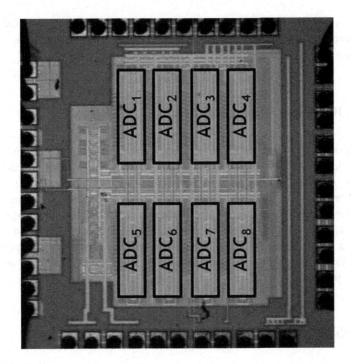

Fig. 6.2 Die photo

6.1.3 Data Capture Cards

Two data capture cards were used with the PCB. Both cards communicated with the computer through a USB interface. The first data capture card was the TSW 1200 [70] that had an LVDS receiver which captured the ADC outputs and sent the data to the computer. The second data capture card [71] interfaced with the digital general purpose input/output (GPIO) pins in the DUT, and used CMOS voltage levels. The transmit and receive data rates for this card were much slower than that of the TSW 1200 [70], but its ease of use made it a valuable addition to the system. The computer was able to program the DUT control register and delay line registers via this data card.

6.1.4 Computer

The background timing skew calibration, as presented in Chap. 3, was implemented externally via a computer. The algorithm, implemented in Matlab^{TM}, read in data from the TSW 1200 and updated the skew correction codes required by the delay lines such that timing skew is compensated. These updated codes were then transmitted to the DUT through the GPIO card, which updated the registers controlling the delay lines.

6.2   ADC Measurement Results

This section presents the ADC measurement results. The DNL and INL results for a single sub-ADC are shown, before and after foreground offset calibration. The ADC dynamic performance is shown with and without background timing skew calibration. The *SNDR* is shown as a function of time once the background timing skew calibration is turned on. Furthermore, the ADC's decimated output spectrum is plotted for high frequency input signals, with and without timing skew calibration. The *SNDR* and *SNR* performance as a function of input frequency is shown. Finally, a performance summary and a comparison with other published works are presented.

6.2.1   Static Performance

The DNL and INL were measured by using a low frequency sinusoidal input signal of 10 MHz and collecting the output histogram [72]. This is done for a single sub-ADC, since time-interleaving averages the DNL and INL and results in artificial results [73].

The comparator offset is a limiting factor in the 5 bit flash ADC static performance. However, foreground offset calibration, as discussed in Chap. 5, reduces the comparator offset and improves the DNL and INL. Figures 6.3a and 6.4a show the typical DNL and INL, respectively, of a single sub-ADC before foreground offset calibration. Figures 6.3b and 6.4b show the typical DNL and INL, respectively, after foreground offset calibration, during which each comparator is calibrated and its offset reduced. As is seen from the figures, both the DNL and INL have been reduced to less than ±0.5 LSB, which demonstrates the functionality of the offset calibration scheme.

6.2.2   Timing Skew Calibration

The background timing skew calibration is implemented using the test setup described in Sect. 6.1. The timing skew calibration is turned on, which means that the Matlab^TM program running on the computer reads in the ADC data, calculates

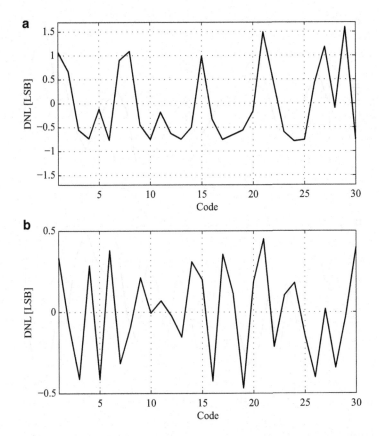

Fig. 6.3 DNL for single sub-ADC (**a**) before offset calibration and (**b**) after offset calibration

the correlation, and updates the delay codes, using algorithms presented in Chap. 3. These delay codes are sent back to the DUT, which updates the registers controlling the delay lines.

The delay codes are updated once every calibration cycle. The time this calibration cycle requires is a function of the calibration clock frequency and the number of samples in each calibration cycle. Different algorithms will need a different number of calibration cycles, depending on how the algorithm is implemented.

In the following results, a calibration clock frequency of 480 MHz, an input signal frequency of approximately 8 GHz, and a sub-ADC clock frequency of 1.5 GHz were used. The ADC output was decimated by a factor of 81, and two different algorithms were implemented. The first set of results was based on the gradient based stochastic maximizer, and the second set was based on the iterative maximizer, both of which are discussed in Chap. 3. As shown in Fig. 6.5a, the SNDR improved from approximately 12 dB to around 24 dB once the calibration was turned on, and converged to a stable point within 20 calibration cycles. In

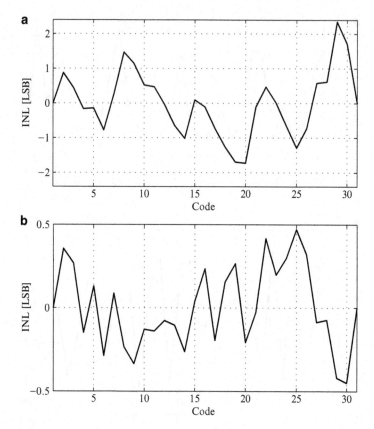

Fig. 6.4 INL for single sub-ADC (**a**) before offset calibration and (**b**) after offset calibration

this example, each calibration cycle consisted of 500,000 samples, which requires approximately 8 ms. This results in a total start up time of approximately 160 ms. Figure 6.5b shows the timing skew calibration codes used by the delay lines and which were updated at the end of each calibration cycle. The change in the delay code for a single delay line is plotted in Fig. 6.6. As is expected, the changes at the beginning of the algorithm are much larger than those once the algorithm converges, as the gradient based stochastic maximizer takes into consideration the gradient of the correlation.

Figure 6.7 shows the SNDR improvement when the iterative maximizer was used. In this example, each calibration cycle consisted of only 50,000 samples. However, over 100 cycles are needed to converge to a stable performance of approximately 24 dB.

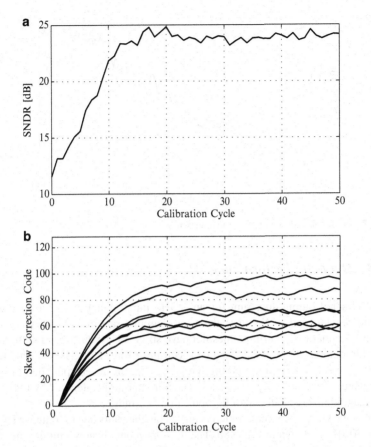

Fig. 6.5 Timing skew calibration algorithm using the gradient based maximizer. (**a**) SNDR convergence and (**b**) timing skew correction codes

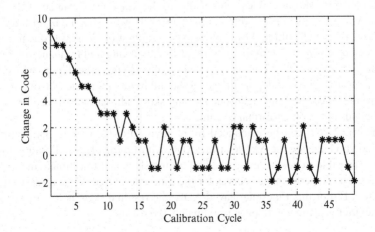

Fig. 6.6 Change in skew correction code after each calibration cycle for a single sub-ADC

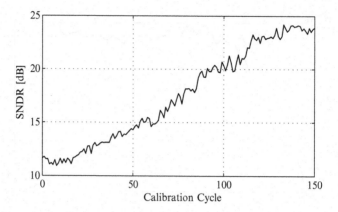

Fig. 6.7 SNDR convergence using iterative maximizer

6.2.3 Dynamic Performance

The decimated output spectrum for an 8 GHz input signal is shown in Fig. 6.8. A frequency larger than Nyquist, given the sampling rate of 12 GS/s, was used to demonstrate that the algorithm does not have strict sub-Nyquist bandwidth limitations. Figure 6.8a shows the decimated spectrum before timing skew calibration is turned on. At this point, the limiting harmonics are the seven spurs due to timing skew, which are denoted by the circles. The third harmonic, denoted by the square, has a magnitude less than that of the spurs due to timing skew.

When the timing skew calibration is turned on, the spurs due to timing skew drop by 10–30 dB, as shown in Fig. 6.8b. The third harmonic is now limiting the SFDR at a magnitude of −31 dBc.

The measured SNDR is plotted as a function of input frequency in Fig. 6.9 with and without background timing skew calibration turned on. At low frequencies, the two curves have similar values. This is due to the low rate of change of the input signal, which results in negligible sampling error. However, as the input frequency increases, the SNDR of the ADC without skew calibration decreases due to timing skew, and results in an approximately 15 dB drop with an input frequency of 8 GHz.

When timing skew calibration is turned on, the *SNDR* curve flattens and suffers only approximately 3 dB degradation over the full frequency range. There is a 12 dB improvement at high frequencies once timing skew calibration is turned on, which corresponds to a 2 bit performance gain.

The ADC *SNR* can be calculated by removing the harmonics, and is plotted in Fig. 6.9b alongside the *SNDR* of the ADC with timing skew calibration turned on. It is possible to calculate the residual timing skew and to estimate the thermal jitter from these performance curves for the time-interleaved ADC, as described in Appendix E. For this time-interleaved ADC, the residual skew is less than 0.4 ps, and the jitter is estimated to be approximately 0.6 psrms.

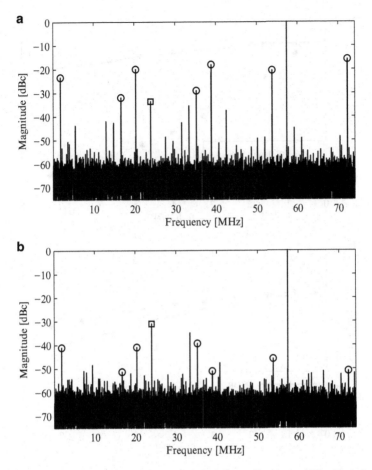

Fig. 6.8 Decimated output spectrum (**a**) without timing skew calibration and (**b**) with timing skew calibration

6.2.4 Performance Summary

The ADC performance is summarized in Table 6.2. The main characteristics of this ADC is that it is implemented in a TSMC 65 nm GP process, runs with a 1.1 V supply, and has a sample rate of 12 GS/s. It has a full scale range of 590 mV. The SNDR at Nyquist is 25.1 dB. The Walden figure-of-merit (FOM) [74], which is calculated with

$$FOM = \frac{P}{f_s \cdot 2^{ENOB}} \tag{6.1}$$

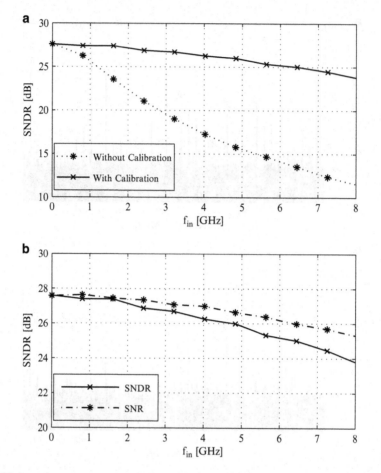

Fig. 6.9 Input frequency sweep. (**a**) *SNDR* performance with and without calibration. (**b**) *SNR* and *SNDR* curves with calibration

is 0.35 pJ/conv-step and 0.46 pJ/conv-step for low and high input frequencies, respectively. The power consumption of the time-interleaved ADC, excluding the digital backend, input/output cells, and the input clock buffer, is 81 mW.

6.2.5 Comparisons

The measured data allows a comparison with other published ADCs by plotting the ADC energy, calculated with P/f_s, versus the SNDR. Figure 6.10a plots the results of ADCs published at the International Solid-State Circuits Conference (ISSCC) and the VLSI Circuit Symposium since 1997 [80] with a sample rate of

Table 6.2 Performance summary of prototype ADC

Parameter	Value	
Process	TSMC 65 nm GP	
Active area	0.44 mm^2	
VDD	1.1 V	
Full scale range	590 mV	
Resolution	5 b	
Sample rate	12 GS/s	
	f_{in} = 10 MHz	f_{in} = 6 GHz
SNDR	27.5 dB	25.1 dB
FOM	0.35 pJ/conv-step	0.46 pJ/conv-step
Power	81 mW (excluding digital backend, I/O cells, and input clock buffer)	

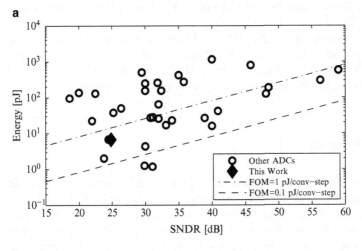

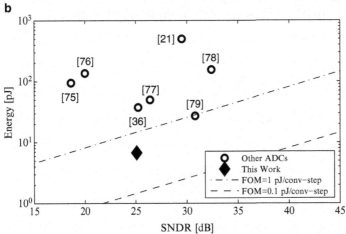

Fig. 6.10 Comparisons between ADCs with a sample rate larger than (**a**) 1 GS/s and (**b**) 10 GS/s

Table 6.3 Published ADCs faster than 10 GS/s

Reference	Resolution [Bits]	Sample rate [GS/s]	Power [W]	SNDR [dB]	Technology
[21]	8	20	9	29.5	0.18 μm
[36]	6	40	1.5	25.2	65 nm
[75]	3	40	3.8	18.6	SiGe
[76]	5	22	3	20	SiGe
[77]	6	24	1.2	26.4	90 nm
[78]	6	10.3	1.6	32.4	90 nm
[79]	6	16	0.435	30.8	65 nm
This work	5	12	0.081	25.1	65 nm

more than 1 GS/s. For a given SNDR, ADCs with lower energy are more efficient. The two lines denote the boundary of ADCs with 1 pJ/conv-step and 0.1 pJ/conv-step FOMs.

If the sample rate of the ADCs is limited to at least 10 GS/s, the comparison consists of only a handful of ADCs. These are tabulated in Table 6.3 and are plotted in Fig. 6.10b. Although the prototype ADC is not the fastest ADC, it is the most power efficient ADC published to date operating above 10 GS/s.

6.3   Summary

In this section, the test setup used to gather measurement results was described. The measurement results were then presented, including DNL and INL plots before and after foreground offset calibration, convergence plots for the background timing skew calibration, and dynamic performance metrics including the output spectrum and SNDR curves. When compared to other published ADCs with sample rates larger than 10 GS/s, the designed ADC is the most power-efficient.

Chapter 7
Conclusion

7.1 Summary

A digitally-equalized serial link uses the digital domain to implement some of the required equalization blocks, which necessitates the use of an ADC. The specifications for such an ADC typically require a time-interleaved ADC be used. This architecture, however, suffers from time-varying errors, which degrade the performance. The relationships between these errors and the performance degradation were detailed in Chap. 2.

Of the main errors in time-interleaved ADCs, timing skew is the most prominent as its effect increases with input frequency. With the high-input signal bandwidth in communication systems, the resulting sub-picosecond constraint on timing skew is extremely difficult to achieve due to all the sources of timing errors in the clock and signal path. Mitigating the effect of timing skew is important such that the dynamic performance specifications of the time-interleaved ADC are met. Chapter 3 presented a statistics-based calibration algorithm that calculated the correlation between each sub-ADC and an extra calibration ADC. The obtained information from this correlation is used to adjust a variable delay line, which changes the delay of each sub-ADC clock and compensates for timing skew.

Serial links have tight power bounds, and if the ADC is to be a viable component of the serial link, it must meet these power constraints. Most multi-GS/s ADCs have high power consumption. The prototype ADC fabricated to evaluate the calibration algorithm was designed to minimize power. A high-level optimization framework, which took into consideration the interleaving factor of the ADC, was presented in Chap. 4, and was followed by Chap. 5, which explained the design of all the circuit blocks. In the latter chapter, the comparator offset correction, which allows the transistors be made smaller and thus decreases power consumption, was discussed, and a foreground offset correction algorithm outlined. Using hundreds of calibration DACs, one for each comparator, it was possible to reduce the size of the comparators such that power gains are achieved.

M. El-Chammas and B. Murmann, *Background Calibration of Time-Interleaved Data Converters*, Analog Circuits and Signal Processing, DOI 10.1007/978-1-4614-1511-4_7, © Springer Science+Business Media, LLC 2012

Finally, the prototype ADC was tested, and its static and dynamic performance were shown in Chap. 6. In addition, the calibration algorithm for timing skew was proven to improve performance at high-frequencies. The resulting ADC consumed 81 mW, and is the most power efficient ADC with sample rate larger than 10 GS/s, published to date.

7.2   Future Work

This research can be taken further in several different directions. One avenue is to investigate the use of alternate sub-ADC architectures as opposed to the flash architecture used. The optimization framework changes as a function of the sub-ADC, and thus more optimal corners may be obtained.

Opportunities for future work also exist in adapting the algorithm to comprehensively work for both high- and low-resolution ADCs, and to encompass additional time-varying errors such as offset, gain and bandwidth mismatch.

In this research, the comparator offset correction was implemented in the foreground. Moving this to the background such that the two calibration algorithms for timing skew and comparator offset both run concurrently would further enhance this project, as would providing the calibration ADC with an on-chip clock generator.

In addition, codesigning the ADC along with the rest of the communication system is another possible direction of research. This would entail optimizing the ADC resolution as a function of the equalization algorithms used, which will further reduce the overall power.

The final direction is that of timing. The timing resolution required for time-interleaved ADC can be sub-picosecond, and in some applications, must be less than 100 fs. However, this is past the usual clock jitter created by clock circuitry and delay lines, which currently poses a final barrier on the ADC dynamic performance. Dealing with jitter is imperative if performance limits are to be pushed any further.

Appendix A
Wide-Sense Cyclostationary Signals

For a zero-mean WSCS signal (WSCS), the autocorrelation, denoted by $R(t_1, t_2)$, is periodic with period T_s such that $R(t_1 + T_s, t_2 + T_s) = R(t_1, t_2)$. The ideal sampling phase of the first sub-ADC is denoted by T_0 such that $0 \leq T_0 < T_s$ and

$$y_i[n] = x(nT_s - \tau_i + T_0), \tag{A.1}$$

where $y_i[n]$ is the output of the ith sub-ADC. Following the derivation for WSS signals in Chap. 2, we write

$$e[n] = y[n] - x_o[n]$$

$$= \left(\sum_{i=0}^{N-1} x(nT_s - \tau_i + T_0)\delta_i \right) - \left(\hat{G}x(nT_s - \hat{\tau} + T_0) \right) \tag{A.2}$$

which results in a mean-square error of

$$f\left(\hat{G}, \hat{\tau}\right) = \frac{1}{N} \sum_{i=0}^{N} R(T_0 - \tau_i, T_0 - \tau_i) + \hat{G}^2 R(T_0 - \hat{\tau}, T_0 - \hat{\tau})$$

$$- \frac{2\hat{G}}{N} \sum_{i=0}^{N-1} R(T_0 - \hat{\tau}, T_0 - \tau_i). \tag{A.3}$$

Setting the partial derivative of (A.3) with respect to $\hat{G}$ to 0 results in

$$\hat{G} = \frac{\sum_i R(T_0 - \hat{\tau}, T_0 - \tau_i)}{NR(T_0 - \hat{\tau}, T_0 - \hat{\tau})}. \tag{A.4}$$

M. El-Chammas and B. Murmann, *Background Calibration of Time-Interleaved Data Converters*, Analog Circuits and Signal Processing, DOI 10.1007/978-1-4614-1511-4,
© Springer Science+Business Media, LLC 2012

Replacing (A.4) into (A.3) results in

$$f\left(\hat{G}, \hat{\tau}\right) = \frac{1}{N} \sum_{i=0}^{N-1} R(T_0 - \tau_i, T_0 - \tau_i) - \frac{\left(\sum_{i=0}^{N-1} R(T_0 - \hat{\tau}, T_0 - \tau_i)\right)^2}{NR(T_0 - \hat{\tau}, T_0 - \hat{\tau})} \quad \text{(A.5)}$$

and minimizing (A.5) over $\hat{\tau}$ results in

$$\hat{\tau} = \arg\max_{\tau} \frac{\left(\sum_{i=0}^{N-1} R(T_0 - \tau, T_0 - \tau_i)\right)^2}{NR(T_0 - \tau, T_0 - \tau)} \quad \text{(A.6)}$$

When applied to WSS input signals, (A.4) and (A.6) reduce to (2.38) and (2.40), respectively.

A.1   WSCS Example

The autocorrelation function for the WSCS example in Chap. 2 is derived in this section. Let the transmitted signal be

$$s(t) = \sum_{i=-\infty}^{\lfloor t/T \rfloor} c_i p(t - iT), \quad \text{(A.7)}$$

where $p(t) = u(t) - u(t - T)$ and $c_i \in \{-1, +1\}$. Thus, $s(t)$ is a sum of rectangular waveforms. Furthermore, $R_c(n, m) = \delta_{n-m}$ and $E[c_n] = 0$. If the channel is a first-order low pass filter such that

$$h(t) = e^{-t\omega_{3\,dB}}, \quad \text{(A.8)}$$

then the received signal at the ADC input is

$$x(t) = s(t) * h(t) = \sum_{i=-\infty}^{\lfloor t/T \rfloor} c_i p(t - iT) * h(t) = \sum_{i=-\infty}^{\lfloor t/T \rfloor} c_i f(t - iT), \quad \text{(A.9)}$$

where

$$f(t) = p(t) * h(t) = \begin{cases} 1 - e^{-t\omega_{3\,dB}}, & \text{if } 0 \le t \le T; \\ K^2 e^{-t\omega_{3\,dB}}, & \text{else} \end{cases} \quad \text{(A.10)}$$

with $K = e^{T\omega_{3\,dB}} - 1$.

When $t_1 \in [nT, (n + 1)T)$ and $t_2 \in [(m)T, (m + 1)T)$, define $r = \min(n, m)$, $\hat{t}_1 = t_1 - rT$, and $\hat{t}_2 = t_2 - rT$. Then

$$R(t_1, t_2) = \sum_{i=-\infty}^{n} \sum_{j=-\infty}^{m} E[c_i c_j f(t_1 - iT) f(t_2 - jT)]$$

$$= \sum_{i=-\infty}^{n} \sum_{j=-\infty}^{m} E[c_i c_j] f(t_1 - iT) f(t_2 - jT)$$

$$= \sum_{i=-\infty}^{r} f(t_1 - iT) f(t_2 - iT)$$

$$= \sum_{i=-\infty}^{0} f\left(\hat{t}_1 - iT\right) f\left(\hat{t}_2 - iT\right) \tag{A.11}$$

Replacing $f(t)$ from (A.10) in the above results in

$$R(t_1, t_2) = \begin{cases} \left(1 - e^{-(\hat{t}_1)\omega_{3\,\mathrm{dB}}}\right)\left(1 - e^{-(\hat{t}_2)\omega_{3\,\mathrm{dB}}}\right), & \text{if } m = n \\ \left(1 - e^{-(\hat{t}_1)\omega_{3\,\mathrm{dB}}}\right) K e^{-(\hat{t}_2)\omega_{3\,\mathrm{dB}}}, & \text{if } m > n \\ \left(1 - e^{-(\hat{t}_2)\omega_{3\,\mathrm{dB}}}\right) K e^{-(\hat{t}_1)\omega_{3\,\mathrm{dB}}}, & \text{if } m < n \end{cases}$$

$$+ \sum_{i=-\infty}^{-1} K^2 e^{-(\hat{t}_1 - iT)\omega_{3\,\mathrm{dB}}} e^{-(\hat{t}_2 - iT)\omega_{3\,\mathrm{dB}}}. \tag{A.12}$$

Finally, since

$$\sum_{i=-\infty}^{-1} e^{-(\hat{t}_1 - iT)\omega_{3\,\mathrm{dB}}} e^{-(\hat{t}_2 - iT)\omega_{3\,\mathrm{dB}}} = e^{-(\hat{t}_1 + \hat{t}_2)\omega_{3\,\mathrm{dB}}} \frac{e^{-2T\omega_{3\,\mathrm{dB}}}}{1 - e^{-2T\omega_{3\,\mathrm{dB}}}} \tag{A.13}$$

we have

$$R(t_1, t_2) = \begin{cases} \left(1 - e^{-(\hat{t}_1)\omega_{3\,\mathrm{dB}}}\right)\left(1 - e^{-(\hat{t}_2)\omega_{3\,\mathrm{dB}}}\right), & \text{if } m = n \\ \left(1 - e^{-(\hat{t}_1)\omega_{3\,\mathrm{dB}}}\right) K e^{-(\hat{t}_2)\omega_{3\,\mathrm{dB}}}, & \text{if } m > n \\ \left(1 - e^{-(\hat{t}_2)\omega_{3\,\mathrm{dB}}}\right) K e^{-(\hat{t}_1)\omega_{3\,\mathrm{dB}}}, & \text{if } m < n \end{cases}$$

$$+ K^2 e^{-(\hat{t}_1 + \hat{t}_2)\omega_{3\,\mathrm{dB}}} \frac{e^{-2T\omega_{3\,\mathrm{dB}}}}{1 - e^{-2T\omega_{3\,\mathrm{dB}}}}$$

$$= R(t_1 + T, t_2 + T) \tag{A.14}$$

which is periodic in T. The mean of the signal $x(t)$ is

$$m(t) = E[x(t)] = E\left[\sum_{i=-\infty}^{n} c_i f(t - iT)\right]$$

$$= \sum_{i=-\infty}^{n} E[c_i] f(t - iT)$$

$$= 0$$

$$= m(t + T). \tag{A.15}$$

Thus, this signal is WSCS.

Appendix B
Comparator Power Model

Chapter 4 presents a high-level optimization framework for time-interleaved ADCs and uses a simplified first-order model for a dynamic comparator. The full derivation of the model is presented in this appendix.

Each inverter in the cross-coupled inverter latch in Fig. B.1a consists of a PMOS and NMOS transistor as in Fig. B.1b, each of which has a threshold voltage of V_{tp} and V_{tn}, respectively. The current in each linearized transistor is modeled as a function of the gate and source voltages, such that the current through the PMOS transistor is

$$I_p(t) = \begin{cases} g_{mp} \left(V_{DD} - V_{in}(t) - V_{tp} \right) & \text{if } V_{DD} - V_{in}(t) \geq V_{tp} \text{ and } V_{out}(t) \leq V_{DD} \\ 0 & \text{else} \end{cases}$$

(B.1)

and the current through the NMOS transistor is

$$I_n(t) = \begin{cases} g_{mn} \left(V_{in}(t) - V_{tn} \right) & \text{if } V_{in}(t) \geq V_{tn} \text{ and } V_{out}(t) \geq 0 \\ 0 & \text{else} \end{cases},$$

(B.2)

where g_{mp} and g_{mn} are the transconductance of the NMOS and PMOS transistors, respectively. Thus, once the comparator is strobed and starts regenerating, $I_1(t) = I_{p,1}(t) - I_{n,1}(t)$ and $I_2(t) = I_{p,2}(t) - I_{n,2}(t)$. For simplicity, $V_t = V_{tp} = V_{tn}$ and $g_m = g_{mp} = g_{mn}$, such that the currents flowing into the capacitors are

$$I_1(t) = \begin{cases} g_m \left(V_{DD} - 2V_2(t) \right) & \text{if } V_{DD} - V_t \geq V_2(t) \geq V_t \text{ and } V_{DD} \geq V_1(t) \geq 0 \\ g_m \left(V_{DD} - V_2(t) - V_t \right) & \text{if } V_t \geq V_2(t) \text{ and } V_{DD} \geq V_1(t) \\ g_m \left(V_2(t) - V_t \right) & \text{if } V_2(t) \geq V_{DD} - V_t \text{ and } V_1(t) \geq 0 \\ 0 & \text{else} \end{cases}$$

(B.3)

M. El-Chammas and B. Murmann, *Background Calibration of Time-Interleaved Data Converters*, Analog Circuits and Signal Processing, DOI 10.1007/978-1-4614-1511-4, © Springer Science+Business Media, LLC 2012

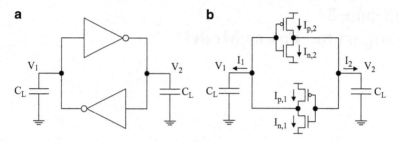

Fig. B.1 Currents in cross-coupled inverter based dynamic latch

and

$$
I_2(t) = \begin{cases} g_m \left(V_{DD} - 2V_1(t) \right) & \text{if } V_{DD} - V_t \geq V_1(t) \geq V_t \text{ and } V_{DD} \geq V_2(t) \geq 0 \\ g_m \left(V_{DD} - V_1(t) - V_t \right) & \text{if } V_t \geq V_1(t) \text{ and } V_{DD} \geq V_2(t) \\ g_m \left(V_1(t) - V_t \right) & \text{if } V_1(t) \geq V_{DD} - V_t \text{ and } V_2(t) \geq 0 \\ 0 & \text{else} \end{cases}
$$

$$(B.4)$$

Both output currents have four regions of operation. The first region is when both the NMOS and PMOS transistors conduct current. Some of this current charges the output capacitor and some is short-circuit current. The second and third region consist of only one of the transistors conducting current such that there is no short-circuit current. In the final region, both transistors are off, since the voltages $V_1(t)$ and $V_2(t)$ have saturated to V_{DD} and ground.

The output voltages $V_1(t)$ and $V_2(t)$ are related to the currents $I_1(t)$ and $I_2(t)$ through the differential equations

$$
I_1(t) = C_L \frac{dV_1(t)}{dt},
$$

$$
I_2(t) = C_L \frac{dV_2(t)}{dt},
$$

$$(B.5)$$

and have initial conditions of

$$
V_1(0) = V_c + v_d/2,
$$

$$
V_2(0) = V_c - v_d/2,
$$

$$(B.6)$$

where V_c is the common-mode voltage and v_d the differential voltage. To simplify the analysis, symmetry is assumed such that $V_c = \frac{V_{DD}}{2}$. Furthermore, $\frac{V_{DD}}{2} - V_t > \frac{v_d}{2} > 0$ such that both transistors in the inverters are on when the comparator is strobed at $t = 0$. Thus, for $t \geq 0$,

$$
g_m \left(V_{DD} - 2V_2(t) \right) = C_L \frac{dV_1(t)}{dt},
$$

$$
g_m \left(V_{DD} - 2V_1(t) \right) = C_L \frac{dV_2(t)}{dt}
$$

$$(B.7)$$

A pair of second-order differential equations can be derived as:

$$\tau_1^2 \frac{d^2 V_1(t)}{dt^2} - V_1(t) + \frac{V_{DD}}{2} = 0,$$

$$\tau_1^2 \frac{d^2 V_2(t)}{dt^2} - V_2(t) + \frac{V_{DD}}{2} = 0, \qquad (B.8)$$

where $\tau_1 = \frac{C_L}{2g_m} = \frac{C_L}{G_m}$. This pair of differential equations has the general solution of

$$V_1(t) = a_1 e^{(-t/\tau)} + b_1 e^{(t/\tau)} + \frac{V_{DD}}{2},$$

$$V_2(t) = a_2 e^{(-t/\tau)} + b_2 e^{(t/\tau)} + \frac{V_{DD}}{2}. \qquad (B.9)$$

As a result of the circuit's initial condition on $V_1(0)$ and $V_2(0)$ as in (B.6), and the additional conditions of

$$\frac{G_m}{2} \cdot (V_{DD} - 2V_2(0)) = C_L \frac{dV_1(0)}{dt}$$

$$\frac{G_m}{2} \cdot (V_{DD} - 2V_1(0)) = C_L \frac{dV_2(0)}{dt} \qquad (B.10)$$

the parameters in (B.9) are derived to be $a_1 = a_2 = 0$ and $b_1 = -b_2 = v_d/2$. Thus,

$$V_t \leq V_1(t) = \frac{v_d}{2} e^{(t/\tau_1)} + \frac{V_{DD}}{2} \leq V_{DD} - V_t,$$

$$V_t \leq V_2(t) = -\frac{v_d}{2} e^{(t/\tau_1)} + \frac{V_{DD}}{2} \leq V_{DD} - V_t. \qquad (B.11)$$

Since $v_d > 0$, $V_1(t)$ increases to $V_{DD} - V_t$ and $V_2(t)$ decreases to V_t. Due to the imposed symmetry, both outputs reach these values at the same time. This phase ends at time

$$t_1 = \tau_1 \cdot \ln \left(\frac{V_{DD} - 2V_t}{v_d} \right). \qquad (B.12)$$

In the second phase of the comparator operation, the PMOS in the first inverter and the NMOS in the second inverter both turn off. Therefore, $I_1(t) = I_p(t)$ and $I_2(t) = -I_n(t)$. Solving the differential equation as before results in

$$V_1(t) = (V_{DD} - 2V_t) e^{\frac{t-t_1}{\tau_2}} + V_t \leq V_{DD}$$

$$0 \leq V_2(t) = -(V_{DD} - 2V_t) e^{\frac{t-t_1}{\tau_2}} + (V_{DD} - V_t) \qquad (B.13)$$

for $t > t_1$ and where $\tau_2 = 2\tau_1$. The required regeneration time for the comparator is the time needed for the outputs to reach V_{DD} and 0, and is

$$T_r = t_1 + \tau_2 \cdot \ln\left(\frac{V_{DD} - V_t}{V_{DD} - 2V_t}\right) = \tau_1 \cdot \ln\left(\frac{(V_{DD} - V_t)^2}{v_d \cdot (V_{DD} - 2V_t)}\right). \tag{B.14}$$

The total current going through V_{DD} is

$$I_{V_{DD}}(t) = \begin{cases} \frac{G_m}{2} \cdot (V_{DD} - 2V_t) & \text{if } 0 \le t \le t_1 \\ \frac{G_m}{2} \cdot \left((V_{DD} - 2V_t) e^{\frac{t-t_1}{\tau_2}}\right) & \text{if } t_1 < t \le T_r. \\ 0 & \text{else} \end{cases} \tag{B.15}$$

The power dissipation results from the current drawn through the power supply, and is

$$P_{comp} = \frac{1}{T_s} \int_0^{T_s} V_{DD} I_{V_{DD}}(t) dt. \tag{B.16}$$

Using (B.15) results in

$$
\begin{aligned}
P_{comp} &= \frac{1}{T_s} \int_0^{t_1} V_{DD} I_{V_{DD}}(t) dt + \frac{1}{T_s} \int_{t_1}^{T_r} V_{DD} I_{V_{DD}}(t) dt \\
&= \frac{G_m}{2} \cdot \frac{V_{DD} \cdot (V_{DD} - 2V_t)}{T_s} \cdot \left[(t_1) + \tau_2 \left(e^{\frac{T_r - t_1}{\tau_2}} - 1\right)\right] \\
&= \frac{\tau_1 G_m}{2} \cdot \frac{V_{DD} \cdot (V_{DD} - 2V_t)}{T_s} \left[\ln\left(\frac{V_{DD} - 2V_t}{v_d}\right) + 2\frac{V_t}{V_{DD} - 2V_t}\right] \\
&= \frac{C_L}{2} \cdot \frac{V_{DD} \cdot (V_{DD} - 2V_t)}{T_s} \cdot \ln\left(\frac{V_{DD} - 2V_t}{v_d}\right) + C_L \cdot \frac{V_{DD} \cdot V_t}{T_s}.
\end{aligned} \tag{B.17}
$$

The power dissipated is divided into two parts. The first coincides with the power in the first phase of the comparator operation, and is a function of v_d. The smaller v_d is, the more short-circuit current is conducted. The second part coincides to the scenario in which one of the transistors is turned off, in which case all the current drawn from the power supply is used to charge the capacitor, which results in the standard dynamic power consumption equation.

In Chap. 4 these equations are used to present a high-level optimization framework. For simplicity, the threshold voltage is set to 0, in which case both transistors in the inverters always conduct current, and the second part of the power equation disappears. Therefore, the output voltages becomes

$$0 \le V_1(t) = \frac{v_d}{2} e^{(t/\tau_1)} + \frac{V_{DD}}{2} \le V_{DD}$$

$$0 \le V_2(t) = -\frac{v_d}{2} e^{(t/\tau_1)} + \frac{V_{DD}}{2} \le V_{DD} \tag{B.18}$$

and the current through the power supply is

$$I_{V_{DD}}(t) = \begin{cases} \frac{G_m}{2} \cdot (V_{DD}) & \text{if } 0 \le t \le T_r \\ 0 & \text{else} \end{cases}. \tag{B.19}$$

The dissipated power becomes

$$P_{comp} = \frac{V_{DD}^2}{T_s} \cdot \frac{C_L}{2} \cdot \ln\left(\frac{V_{DD}}{v_d}\right). \tag{B.20}$$

The comparator power in (B.20) is proportional to both V_{DD} and C_L, and inversely proportional to the sampling period T_s and the differential input voltage v_d.

Appendix C
Optimizing a Transistor-Level Comparator

In Sect. 4.3, the high-level optimization framework presented in Chap. 4 was extended to a transistor-level circuit. This appendix chapter elaborates on the plotted results and explains the simulation setup.

The data presented in Sect. 4.3 was based on the comparator explained in Sect. 5.1.2 and shown in Fig. C.1. The input voltages V_{inp} and V_{inn} have a common-mode voltage of V_c and a differential input voltage of v_d. The following parameters are used in the comprehensive search: the width of the input transistors M_{1-4}, the width of the clock transistor M_{clk}, the widths of the kickback transistors M_{KB1} and M_{KB2}, the widths of the inverter transistors, and the common-mode voltage V_c. A supply voltage of 1 V is used.

A Perl script was written to create a large number of OCEAN files, each of which has a different set of parameters. The script took in the following parameters:

```
for i1=1:8
  for i2=1:5
    for i3=1:3
      for i4=1:5
        for i5=1:5
          for i6 = 1:3
              Winvn = 1e-6*(1+0.5*i1)
              Win = Winvn*(0.5*i2+1)
              Winvp = Winvn*(0.5*i3+1)
              Wclk =  Winvn*(0.5*i4+0.5)
              Wkb= Winvn*(0.5*i5+1)
              Vc = 0.55+0.05*i6
          end
        end
      end
    end
  end
end
```

M. El-Chammas and B. Murmann, *Background Calibration of Time-Interleaved Data Converters*, Analog Circuits and Signal Processing, DOI 10.1007/978-1-4614-1511-4, © Springer Science+Business Media, LLC 2012

Fig. C.1 Schematic of
dynamic comparator

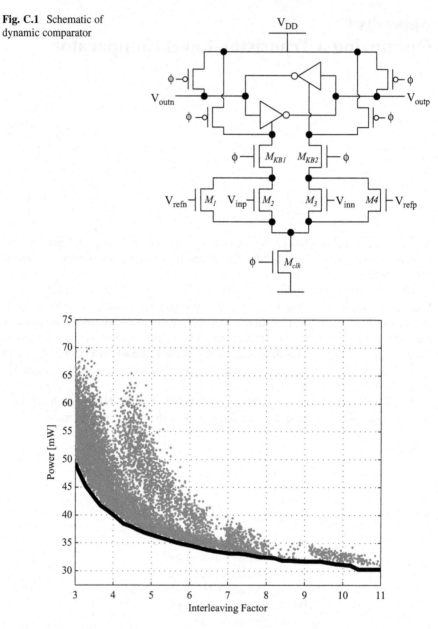

Fig. C.2 Simulated time-interleaved ADC power with different transistor sizings. The optimal
boundary is outlined in black

A total of 9,000 OCEAN scripts were then consecutively run. With more
computing power available, these scripts could be run simultaneously to speed
up the data collection. At the end of each simulation, the power dissipation was

calculated, as well as the time delay from the rising edge of the clock, at which point the comparator starts regenerating, to the point at which the output differential voltage reaches $0.95V_{DD}$. This delay t_d is used to calculate the minimum possible sub-ADC sampling period, which is equal to $2t_d$, from which the interleaving factor, for a given time-interleaved ADC sample rate, can be calculated. The data is then aggregated in Fig. C.2, as discussed in Sect. 4.3. Although this plot shows all feasible combinations, the optimal curve, which is outlined in black, consists of the comparator realizations that have minimum power dissipation for a given interleaving factor.

Appendix D
Comparator Skew

The comparators in a flash ADC ideally sample the input signal at the same instance. If there is skew between the latching points of the comparator, which can result from the clock distribution network and from the comparator transistor variations, then each comparator samples the input signal at a slightly different time, as shown with the clock timing diagrams in Fig. D.1.

The digital output of this bank of M comparators is written as a sum of the comparator outputs, assuming a ones-adder (also referred to as a Wallace Encoder.) Without skew, the output at time nT_s is

$$D_{out}[n] = \sum_{i=1}^{M} \text{sign}(v_{in}(nT_s) - v_{r,i}), \tag{D.1}$$

where $v_{in}(t)$ is the input signal, $v_{r,i}$ is the reference voltage for the ith comparator, and T_s is the sampling period. With skew, this output becomes

$$D_{out}[n] = \sum_{i=1}^{M} \text{sign}(v_{in}(nT_s + \alpha_i) - v_{r,i}), \tag{D.2}$$

where α_i is the skew for the ith comparator.

For small values of α_i, the input signal can be approximated with its Taylor series expansion as:

$$v_{in}(nT_s + \alpha_i) \approx v_{in}(nT_s) + \alpha_i \cdot v'_{in}(nT_s) \tag{D.3}$$

with the assumption that $\alpha_i \ll 1$ for all i.

Thus,

$$D_{out}[n] \approx \sum_{i=1}^{M} \text{sign}(\hat{v}_{in}(nT_s) - v_{r,i}), \tag{D.4}$$

M. El-Chammas and B. Murmann, *Background Calibration of Time-Interleaved Data Converters*, Analog Circuits and Signal Processing, DOI 10.1007/978-1-4614-1511-4, © Springer Science+Business Media, LLC 2012

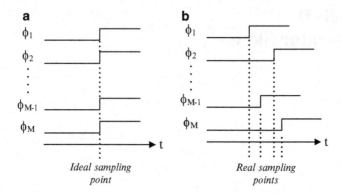

Fig. D.1 Comparator clock sampling edges (**a**) without skew and (**b**) with skew

where

$$\hat{v}_{in}(nT_s) = v_{in}(nT_s) + \alpha_i \cdot v'_{in}(nT_s). \tag{D.5}$$

The signal $\hat{v}_{in}(nT_s)$ can be viewed as a noisy version of the input signal. The noise in this signal is represented by $v_{no}[n] = \alpha_i \cdot v'_{in}(nT_s)$, which was the second term of the Taylor expansion in (D.3). Assuming that α_i is independent and identically distributed, this can be represented by its mean and variance at the nth time sample such that

$$m_{no}[n] = E\left[v_{no}[n]\right] = v'_{in}(nT_s) \cdot \left(\frac{1}{M}\sum_{i=1}^{M}\alpha_i\right) \tag{D.6}$$

and

$$
\begin{aligned}
\sigma_{no}^2[n] &= E[(v_{no}[n] - m_{no}[n])^2] \\
&= \frac{1}{M}\sum_{i=1}^{M}\left(v'_{in}(nT_s)\right)^2\alpha_i^2 - \left(v'_{in}(nT_s)\right)^2\frac{1}{M^2}\left(\sum_{i=1}^{M}\alpha_i\right)^2 \\
&= \frac{\left(v'_{in}(nT_s)\right)^2}{M}\cdot\left(\sum_{i=1}^{M}\alpha_i^2 - \frac{1}{M}\left(\sum_{i=1}^{M}\alpha_i\right)^2\right) \\
&= \frac{\left(v'_{in}(nT_s)\right)^2}{M}\cdot\sum_{i=1}^{M}\left(\frac{M-1}{M}\right)\alpha_i^2.
\end{aligned}
\tag{D.7}
$$

Both the variance and mean at time sample n are a function of the derivative of the input signal at that time sample. Taking the average of (D.7) over n results in

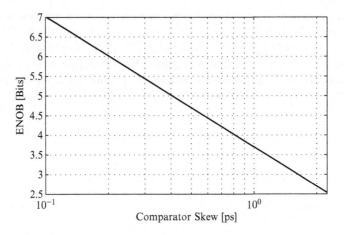

Fig. D.2 ADC ENOB as a function of the comparator skew

$$\bar{\sigma}_{no}^2 = E\left[\sigma_{no}^2[n]\right]$$

$$= \frac{E[(v'_{in}(nT_s))^2]}{M} \cdot \sum_{i=1}^{M}\left(\frac{M-1}{M}\right) E[\alpha_i^2]$$

$$= E[(v'_{in}(nT_s))^2] \cdot \left(\frac{M-1}{M}\right) \cdot \sigma_\alpha^2, \tag{D.8}$$

where σ_α is the variance of the comparator skew. Thus, for slow signals, which have smaller signal derivative, the variance in (D.8) is smaller than for fast signals. This is intuitive due to the slower rate of change in the signal, which results in a smaller voltage error for a given time difference.

For an input signal with power P and an ADC quantization noise variance of σ_q^2, the resulting *SNR* including the effect of comparator skew is

$$SNR = \frac{P}{\sigma_q^2 + \bar{\sigma}_{no}^2}. \tag{D.9}$$

As an example, assume the input signal is a sinusoidal function with frequency f_{in}, such that, $v_{in}(nT_s) = A\sin(2\pi f_{in}(nT_s))$ and $v'_{in}(nT_s) = 2\pi f_{in}A\cos(2\pi f_{in}(nT_s))$. This results in

$$E[(v'_{in}(nT_s))^2] = (2\pi f_{in})^2 \cdot \frac{A^2}{2}. \tag{D.10}$$

To simplify the calculation of (D.8), and without much loss of generality, assume an infinite resolution ADC is used such that $M = \infty$ and $\sigma_q = 0$. Thus, the *SNR* is

$$SNR = \frac{1}{(2\pi f_{in}\sigma_\alpha)^2} \tag{D.11}$$

which decreases both as a function of input frequency and as a function of the skew variance. For a 10 GHz input signal, the resulting resolution of the ADC is shown in Fig. D.2. With only 2 ps of comparator skew, the ADC effective number of bits, calculated with $\frac{SNR-1.76}{6.02}$, has dropped below 3 bits. Therefore, using a track-and-hold with high-speed input signals is important, as it holds the input signal, such that comparator skew has negligible effect.

Appendix E
Calculating Residual Timing Errors

This appendix details the method used in Chap. 6 to calculate the residual timing skew and the estimated jitter of the prototype ADC.

E.1 Residual Timing Skew

Given the decimated output spectrum obtained with an input sinusoidal signal, the positions of the timing skew spurs are known, as derived in Chap. 2 and in [6]. With an interleaving factor of N, there are $N - 1$ spurs to account for. The magnitude of these spurs is $A[k]$, for $k = 1, \ldots, N - 1$. Theoretically, the value of $A[k]$ is

$$A[k] = \frac{1}{N} \sum_{i=0}^{N-1} e^{-j 2\pi \tau_i f_{in}} \cdot e^{-j \frac{2ki\pi}{N}}. \qquad (E.1)$$

This can be simplified into $A = BC$, where B is a matrix such that

$$B[i,k] = \frac{1}{N} \cdot e^{-j \frac{2ki\pi}{N}} \qquad (E.2)$$

and C is a vector such that

$$C[i] = e^{-j 2\pi \tau_i f_{in}}. \qquad (E.3)$$

Thus,

$$A[k] = \sum_{i=0}^{N-1} B[i,k] \cdot C[i]. \qquad (E.4)$$

The vector A is known from the measured data, and the matrix B is a function of N. Thus, $C = B^{-1} A$, where B^{-1} is the pseudoinverse of B. The vector of timing skews is then calculated by:

M. El-Chammas and B. Murmann, *Background Calibration of Time-Interleaved Data Converters*, Analog Circuits and Signal Processing, DOI 10.1007/978-1-4614-1511-4, © Springer Science+Business Media, LLC 2012

$$\tau = \frac{\ln(C)}{-j2\pi f_{\text{in}}}.$$ (E.5)

Finally, the residual timing skew is

$$\sigma_\tau = \sqrt{\frac{\sum_{i=0}^{N-1} \tau_i^2}{N}}.$$ (E.6)

This provides a simple approach to calculate the residual timing skew from measured data.

E.2   Estimated Jitter

The systematic timing errors of timing skew was presented in the previous section. A second timing error is that due to random jitter. To estimate the random jitter from the measured output spectrum, higher-order harmonics are removed from the ADC output spectrum, which is possible since their position is a function of the position of the input signal fundamental tone. The remaining performance limitations come from quantization and thermal noise, jitter, and timing skew. The degradation due to quantization and thermal noise can be calculated from low frequency input signals, since jitter and timing skew have a negligible effect at those frequencies. Assuming that quantization and thermal noise do not increase as a function of input frequency, any degradation in *SNR* with increase in frequency is due only to jitter and timing skew.

The effect of timing skew can either be calculated using the approach outlined in the preceding section, or it can be removed by simply deleting the spurs due to timing skew. A problem with the second approach is that spurs due to timing skew arise as a function of both the fundamental input tone and of higher-order harmonics, so care has to be taken to ensure that all tones are removed.

If the residual skew is estimated as in Sect. E.1, then the degradation due to both jitter and timing skew with an input sinusoidal signal is approximately

$$\sigma_T^2 \approx P \cdot (2\pi f_{in})^2 \cdot \left(\sigma_\tau^2 + \sigma_j^2\right),$$ (E.7)

where σ_τ is as calculated in (E.6). Thus,

$$SNR = \frac{P}{\sigma_q^2 + \sigma_T^2},$$ (E.8)

where σ_q^2 is the combined variance of quantization and thermal noise. The value of σ_j that equates the *SNR* in (E.8) to the measured *SNR* with high-frequency inputs is the estimated jitter.

References

1. STA, "Serial Attached SCSI - Roadmap," http://www.scsita.org/sas_library/sas-master-roadmaps/.
2. V. Balan, J. Caroselli, J.-G. Chern, C. Chow, R. Dadi, C. Desai, L. Fang, D. Hsu, P. Joshi, H. Kimura, C. Liu, T.-W. Pan, R. Park, C. You, Y. Zeng, E. Zhang, and F. Zhong, "A 4.8-6. 4-Gb/s Serial Link for Backplane Applications Using Decision Feedback Equalization," *IEEE Journal of Solid-State Circuits*, vol. 40, no. 9, pp. 1957–1967, Sept. 2005.
3. M. Harwood, N. Warke, R. Simpson, T. Leslie, A. Amerasekera, S. Batty, D. Colman, E. Carr, V. Gopinathan, S. Hubbins, P. Hunt, A. Joy, P. Khandelwal, B. Killips, T. Krause, S. Lytollis, A. Pickering, M. Saxton, D. Sebastio, G. Swanson, A. Szczepanek, T. Ward, J. Williams, R. Williams, and T. Willwerth, "A 12.5Gb/s SerDes in 65nm CMOS Using a Baud-Rate ADC with Digital Receiver Equalization and Clock Recovery," in *Proceedings of IEEE International Solid-State Circuits Conference*, vol. 1, Feb. 2007, pp. 436–591.
4. H. Chung, A. Rylyakov, Z. T. Deniz, J. Bulzacchelli, G.-Y. Wei, and D. Friedman, "A 7.5-GS/s 3.8-ENOB 52-mW Flash ADC with Clock Duty Cycle Control in 65nm CMOS," in *VLSI Circuits Symposium, Digest of Technical Papers*, June 2009, pp. 268–269.
5. W. Black and D. Hodges, "Time Interleaved Converter Arrays," *IEEE Journal of Solid-State Circuits*, vol. 15, no. 6, pp. 1022–1029, Dec. 1980.
6. Y.-C. Jenq, "Digital Spectra of Nonuniformly Sampled Signals: Fundamentals and High-Speed Waveform Digitizers," *IEEE Transactions on Instrumentation and Measurement*, vol. 37, no. 2, pp. 245–251, June 1988.
7. J. Proakis and D. Manolakis, *Digital Signal Processing*, 3rd ed. Prentice Hall, New Jersey, USA, 1996.
8. A. V. Oppenheim, A. S. Willsky, and S. Hamid, *Signals and Systems*, 2nd ed. Prentice Hall, New Jersey, USA, 1996.
9. N. Kurosawa, H. Kobayashi, K. Maruyama, H. Sugawara, and K. Kobayashi, "Explicit Analysis of Channel Mismatch Effects in Time-Interleaved ADC Systems," *IEEE Transactions on Circuits and Systems I: Fundamental Theory and Applications*, vol. 48, no. 3, pp. 261–271, March 2001.
10. M. El-Chammas and B. Murmann, "General Analysis on the Impact of Phase-Skew in Time-Interleaved ADCs," *IEEE Transactions on Circuits and Systems I: Regular Papers*, vol. 56, no. 5, pp. 902–910, May 2009.
11. A. Papoulis, *Probability, Random Variables, and Stochastic Processes*, 3rd ed. McGraw-Hill, New York, 1991.
12. C. Vogel, "The Impact of Combined Channel Mismatch Effects in Time-Interleaved ADCs," *IEEE Transactions on Instrumentation and Measurement*, vol. 54, no. 1, pp. 415–427, Feb. 2005.

M. El-Chammas and B. Murmann, *Background Calibration of Time-Interleaved Data Converters*, Analog Circuits and Signal Processing, DOI 10.1007/978-1-4614-1511-4, © Springer Science+Business Media, LLC 2012

13. N. Da Dalt, M. Harteneck, C. Sandner, and A. Wiesbauer, "On the Jitter Requirements of the Sampling Clock for Analog-to-Digital Converters," *IEEE Transactions on Circuits and Systems I: Fundamental Theory and Applications*, vol. 49, no. 9, pp. 1354–1360, Sept. 2002.

14. S. Louwsma, A. van Tuijl, M. Vertregt, and B. Nauta, "A 1.35 GS/s, 10 b, 175 mW Time-Interleaved AD Converter in 0.13 μm CMOS," *IEEE Journal of Solid-State Circuits*, vol. 43, no. 4, pp. 778–786, Apr. 2008.

15. M. Pelgrom, A. Duinmaijer, and A. Welbers, "Matching Properties of MOS Transistors," *IEEE Journal of Solid-State Circuits*, vol. 24, no. 5, pp. 1433–1439, Oct. 1989.

16. A. Agrawal, A. Liu, P. Hanumolu, and G.-Y. Wei, "An 8 × 5 Gb/s Parallel Receiver With Collaborative Timing Recovery," *IEEE Journal of Solid-State Circuits*, vol. 44, no. 11, pp. 3120–3130, Nov. 2009.

17. S. Gupta, M. Inerfield, and J. Wang, "A 1-GS/s 11-bit ADC with 55-dB SNDR, 250-mW Power Realized by a High Bandwidth Scalable Time-Interleaved Architecture," *IEEE Journal of Solid-State Circuits*, vol. 41, no. 12, pp. 2650–2657, Dec. 2006.

18. K. Poulton, J. Corcoran, and T. Hornak, "A 1-GHz 6-bit ADC System," *IEEE Journal of Solid-State Circuits*, vol. 22, no. 6, pp. 962–970, Dec. 1987.

19. S. Jamal, D. Fu, N.-J. Chang, P. Hurst, and S. Lewis, "A 10-b 120-Msample/s Time-Interleaved Analog-to-Digital Converter with Digital Background Calibration," *IEEE Journal of Solid-State Circuits*, vol. 37, no. 12, pp. 1618–1627, Dec. 2002.

20. T. Laakso, V. Valimaki, M. Karjalainen, and U. Laine, "Splitting the Unit Delay - Tools for Fractional Delay Filter Design," *IEEE Signal Processing Magazine*, vol. 13, no. 1, pp. 30–60, Jan. 1996.

21. K. Poulton, R. Neff, B. Setterberg, B. Wuppermann, T. Kopley, R. Jewett, J. Pernillo, C. Tan, and A. Montijo, "A 20 GS/s 8 b ADC with a 1 MB Memory in 0.18 μm CMOS," in *Proceedings of IEEE International Solid-State Circuits Conference*, vol. 1, Feb. 2003, pp. 318–496.

22. D. Camarero, K. Ben Kalaia, J.-F. Naviner, and P. Loumeau, "Mixed-Signal Clock-Skew Calibration Technique for Time-Interleaved ADCs," *IEEE Transactions on Circuits and Systems I: Regular Papers*, vol. 55, no. 11, pp. 3676–3687, Dec. 2008.

23. S. Boyd and L. Vandendberghe, *Convex Optimization*. Cambridge University Press, New York, 2004.

24. W. Root and W. Davenport, *An Introduction to the Theory of Random Signals and Noise*, 2nd ed. McGraw-Hill, New York, 1958.

25. J. J. Bussgang, "Crosscorrelation Functions of Amplitude-Distorted Gaussian Signals," *MIT Research Laboratory of Electronics Technical Reports*, no. 216, March 1952.

26. C.-Y. Wang and J.-T. Wu, "A Multiphase Timing-Skew Calibration Technique Using Zero-Crossing Detection," *IEEE Transactions on Circuits and Systems I: Regular Papers*, vol. 56, no. 6, pp. 1102–1114, June 2009.

27. J. Van Vleck and D. Middleton, "The Spectrum of Clipped Noise," *Proceedings of the IEEE*, vol. 54, no. 1, pp. 2–19, Jan. 1966.

28. F. Todero, "On Some Bonds Between Autocorrelation and Power Spectra Functions," *Proceedings of the IEEE*, vol. 56, no. 12, pp. 2170–2171, Dec. 1968.

29. J. McFadden, "The Axis-Crossing Intervals of Random Functions," *IRE Transactions on Information Theory*, vol. 2, no. 4, pp. 146–150, Dec. 1956.

30. H. Pan and A. Abidi, "Signal Folding in A/D Converters," *IEEE Transactions on Circuits and Systems I: Regular Papers*, vol. 51, no. 1, pp. 3–14, Jan. 2004.

31. R. Van De Plassche and R. Van Der Grift, "A High-Speed 7 Bit A/D Converter," *IEEE Journal of Solid-State Circuits*, vol. 14, no. 6, pp. 938–943, Dec. 1979.

32. T. Toifl, C. Menolfi, M. Ruegg, R. Reutemann, P. Buchmann, M. Kossel, T. Morf, J. Weiss, and M. Schmatz, "A 22-Gb/s PAM-4 Receiver in 90-nm CMOS SOI Technology," *IEEE Journal of Solid-State Circuits*, vol. 41, no. 4, pp. 954–965, Apr. 2006.

33. T. Toifl, "Design Techniques for Ultra-Low-Power and Compact Transceivers in CMOS," in *Proceedings of IEEE International Symposium of Solid-State Circuits Conference 2008, ATAC Design Forum: Future of High-Speed Transceivers*, Feb. 2008.

34. A. Nikoozadeh and B. Murmann, "An Analysis of Latch Comparator Offset Due to Load Capacitor Mismatch," *IEEE Transactions on Circuits and Systems II: Express Briefs*, vol. 53, no. 12, pp. 1398–1402, Dec. 2006.
35. H. Veendrick, "The Behaviour of Flip-Flops Used as Synchronizers and Prediction of their Failure Rate," *IEEE Journal of Solid-State Circuits*, vol. 15, no. 2, pp. 169–176, Apr. 1980.
36. Y. Greshishchev, J. Aguirre, M. Besson, R. Gibbins, C. Falt, P. Flemke, N. Ben-Hamida, D. Pollex, P. Schvan, and S.-C. Wang, "A 40GS/s 6b ADC in 65nm CMOS," in *Proceedings of IEEE International Solid-State Circuits Conference*, vol. 1, Feb. 2010, pp. 390–391.
37. G. Van der Plas, S. Decoutere, and S. Donnay, "A 0.16pJ/Conversion-Step 2.5mW 1.25GS/s 4b ADC in a 90nm Digital CMOS Process," in *Proceedings of IEEE International Solid-State Circuits Conference*, vol. 1, Feb. 2006, p. 2310.
38. W. Yu, S. Sen, and B. Leung, "Distortion Analysis of MOS Track-and-Hold Sampling Mixers using Time-Varying Volterra Series," *IEEE Transactions on Circuits and Systems II: Analog and Digital Signal Processing*, vol. 46, no. 2, pp. 101–113, Feb. 1999.
39. J. Steensgaard, "Bootstrapped Low-Voltage Analog Switches," in *Proceedings of IEEE International Symposium on Circuits and Systems*, vol. 2, July 1999, pp. 29–32.
40. A. Abo and P. Gray, "A 1.5-V, 10-bit, 14.3-MS/s CMOS Pipeline Analog-to-Digital Converter," *IEEE Journal of Solid-State Circuits*, vol. 34, no. 5, pp. 599–606, May 1999.
41. J. B. Johnson, "Thermal Agitation of Electricity in Conductors," *Phys. Rev.*, vol. 32, no. 1, p. 97, July 1928.
42. H. Nyquist, "Thermal Agitation of Electric Charge in Conductors," *Phys. Rev.*, vol. 32, no. 1, pp. 110–113, July 1928.
43. T. Kobayashi, K. Nogami, T. Shirotori, and Y. Fujimoto, "A Current-Controlled Latch Sense Amplifier and a Static Power-Saving Input Buffer for Low-Power Architecture," *IEEE Journal of Solid-State Circuits*, vol. 28, no. 4, pp. 523–527, Apr. 1993.
44. C. Portmann and T. Meng, "Power-Efficient Metastability Error Reduction in CMOS Flash A/D Converters," *IEEE Journal of Solid-State Circuits*, vol. 31, no. 8, pp. 1132–1140, Aug. 1996.
45. P. Nuzzo, F. De Bernardinis, P. Terreni, and G. Van der Plas, "Noise Analysis of Regenerative Comparators for Reconfigurable ADC Architectures," *IEEE Transactions on Circuits and Systems I: Regular Papers*, vol. 55, no. 6, pp. 1441–1454, July 2008.
46. J. Kim, B. Leibowitz, J. Ren, and C. Madden, "Simulation and Analysis of Random Decision Errors in Clocked Comparators," *IEEE Transactions on Circuits and Systems I: Regular Papers*, vol. 56, no. 8, pp. 1844–1857, Aug. 2009.
47. J. He, S. Zhan, D. Chen, and R. Geiger, "Analyses of Static and Dynamic Random Offset Voltages in Dynamic Comparators," *IEEE Transactions on Circuits and Systems I: Regular Papers*, vol. 56, no. 5, pp. 911–919, May 2009.
48. J. Kim, K. D. Jones, and M. A. Horowitz, "Fast, Non-Monte-Carlo Estimation of Transient Performance Variation Due to Device Mismatch," *IEEE Transactions on Circuits and Systems I: Regular Papers*, vol. 57, no. 7, pp. 1746–1755, July 2010.
49. T. Matthews and P. Heedley, "A Simulation Method for Accurately Determining DC and Dynamic Offsets in Comparators," in *Proceedings of Midwest Symposium on Circuits and Systems*, vol. 2, Aug. 2005, pp. 1815–1818.
50. P. Figueiredo and J. Vital, "Kickback Noise Reduction Techniques for CMOS Latched Comparators," *IEEE Transactions on Circuits and Systems II: Express Briefs*, vol. 53, no. 7, pp. 541–545, July 2006.
51. T. Sundstrom and A. Alvandpour, "A Kick-Back Reduced Comparator for a 4-6-Bit 3-GS/s Flash ADC in a 90nm CMOS Process," in *Proceedings of International Conference on Mixed Design of Integrated Circuits and Systems*, June 2007, pp. 195–198.
52. P. Nuzzo, G. Van der Plas, R. De Bernardinis, L. Van der Perre, B. Gyselinckx, and P. Terreni, "A 10.6mW/0.8pJ Power-Scalable 1GS/s 4b ADC in 0.18 μm CMOS with 5.8GHz ERBW," in *Proceedings of ACM/IEEE Design Automation Conference*, July 2006, pp. 873–878.
53. S. Park, Y. Palaskas, and M. Flynn, "A 4-GS/s 4-bit Flash ADC in 0.18 μm CMOS," *IEEE Journal of Solid-State Circuits*, vol. 42, no. 9, pp. 1865–1872, Sept. 2007.

54. K.-L. Wong and C.-K. Yang, "Offset Compensation in Comparators with Minimum Input-Referred Supply Noise," *IEEE Journal of Solid-State Circuits*, vol. 39, no. 5, pp. 837–840, May 2004.

55. J. Schoeff, "An Inherently Monotonic 12 Bit DAC," *IEEE Journal of Solid-State Circuits*, vol. 14, no. 6, pp. 904–911, Dec. 1979.

56. L. Samid, P. Volz, and Y. Manoli, "A Dynamic Analysis of a Latched CMOS Comparator," in *Proceedings of IEEE International Symposium on Circuits and Systems*, vol. 1, June 2004, pp. I–181–4.

57. B. Verbruggen, P. Wambacq, M. Kuijk, and G. Van der Plas, "A 7.6 mW 1.75 GS/s 5 Bit Flash A/D converter in 90 nm Digital CMOS," in *VLSI Circuits Symposium, Digest of Technical Papers*, June 2008, pp. 14–15.

58. F. Kaess, R. Kanan, B. Hochet, and M. Declercq, "New Encoding Scheme for High-Speed Flash ADCs," in *Proceedings of IEEE International Symposium on Circuits and Systems*, vol. 1, June 1997, pp. 5–8.

59. E. Sail and M. Vesterbacka, "Thermometer-to-Binary Decoders for Flash Analog-to-Digital Converters," in *Proceedings of European Conference on Circuit Theory and Design*, Aug. 2007, pp. 240–243.

60. C. Paulus, H.-M. Bluthgen, M. Low, E. Sicheneder, N. Bruls, A. Courtois, M. Tiebout, and R. Thewes, "A 4GS/s 6b Flash ADC in 0.13 μm CMOS," in *VLSI Circuits Symposium, Digest of Technical Papers*, June 2004, pp. 420–423.

61. C. S. Wallace, "A Suggestion for a Fast Multiplier," *IEEE Transactions on Electronic Computers*, vol. EC-13, no. 1, pp. 14–17, Feb. 1964.

62. R. Kanan, F. Kaess, and M. Declercq, "A 640 mW High Accuracy 8-Bit 1 GHz Flash ADC Encoder," in *Proceedings of IEEE International Symposium on Circuits and Systems*, vol. 2, June 1999, pp. 420–423.

63. M. Shinagawa, Y. Akazawa, and T. Wakimoto, "Jitter Analysis of High-Speed Sampling Systems," *IEEE Journal of Solid-State Circuits*, vol. 25, no. 1, pp. 220–224, Feb. 1990.

64. A. Abidi, "Phase Noise and Jitter in CMOS Ring Oscillators," *IEEE Journal of Solid-State Circuits*, vol. 41, no. 8, pp. 1803–1816, Aug. 2006.

65. T. Weigandt, B. Kim, and P. Gray, "Analysis of Timing Jitter in CMOS Ring Oscillators," in *Proceedings of IEEE International Symposium on Circuits and Systems*, vol. 4, May 1994, pp. 27–30.

66. A. Strak and H. Tenhunen, "Analysis of Timing Jitter in Inverters Induced by Power-Supply Noise," in *Proceedings of International Conference on Design and Test of Integrated Systems in Nanoscale Technology*, Sept. 2006, pp. 53–56.

67. X. Gao, B. Nauta, and E. Klumperink, "Advantages of Shift Registers Over DLLs for Flexible Low Jitter Multiphase Clock Generation," *IEEE Transactions on Circuits and Systems II: Express Briefs*, vol. 55, no. 3, pp. 244–248, March 2008.

68. A. Boni, A. Pierazzi, and D. Vecchi, "LVDS I/O Interface for Gb/s-per-pin Operation in 0.35 μm CMOS ," *IEEE Journal of Solid-State Circuits*, vol. 36, no. 4, pp. 706–711, Apr. 2001.

69. M. Chen, J. Silva-Martinez, M. Nix, and M. Robinson, "Low-Voltage Low-Power LVDS Drivers," *IEEE Journal of Solid-State Circuits*, vol. 40, no. 2, pp. 472–479, Feb. 2005.

70. Texas Instruments, "TSW 1200 High Speed ADC LVDS Evaluation System," TSW1200 datasheet, Apr. 2007 [Revised Aug. 2008].

71. Nano River Technologies, "Miniboard: USB-I2C/SPI/GPIO Interface Adapter," MiniBoard user guide, Apr. 2009.

72. M. Bossche, J. Schoukens, and J. Renneboog, "Dynamic Testing and Diagnostics of A/D Converters," *IEEE Transactions on Circuits and Systems*, vol. 33, no. 8, pp. 775–785, Aug. 1986.

73. J. Simoes, J. Landeck, and C. Correia, "Nonlinearity of a Data-Acquisition System with Interleaving/Multiplexing," *IEEE Transactions on Instrumentation and Measurement*, vol. 46, no. 6, pp. 1274–1279, Dec. 1997.

74. R. Walden, "Performance Trends for Analog to Digital Converters," *IEEE Communications Magazine*, vol. 37, no. 2, pp. 96–101, Feb. 1999.

75. W. Cheng, W. Ali, M.-J. Choi, K. Liu, T. Tat, D. Devendorf, L. Linder, and R. Stevens, "A 3b 40GS/s ADC-DAC in 0.12 μm SiGe," in *Proceedings of IEEE International Solid-State Circuits Conference*, vol. 1, Feb. 2004, pp. 262–263.

76. P. Schvan, D. Pollex, S.-C. Wang, C. Falt, and N. Ben-Hamida, "A 22GS/s 5b ADC in 0.13μm SiGe BiCMOS," in *Proceedings of IEEE International Solid-State Circuits Conference*, vol. 1, Feb. 2006, pp. 2340–2349.

77. P. Schvan, J. Bach, C. Fait, P. Flemke, R. Gibbins, Y. Greshishchev, N. Ben-Hamida, D. Pollex, J. Sitch, S.-C. Wang, and J. Wolczanski, "A 24GS/s 6b ADC in 90nm CMOS," in *Proceedings of IEEE International Solid-State Circuits Conference*, vol. 1, Feb. 2008, pp. 544–634.

78. A. Nazemi, C. Grace, L. Lewyn, B. Kobeissy, O. Agazzi, P. Voois, C. Abidin, G. Eaton, M. Kargar, C. Marquez, S. Ramprasad, F. Bollo, V. Posse, S. Wang, and G. Asmanis, "A 10.3GS/s 6bit (5.1 ENOB at Nyquist) Time-Interleaved Pipelined ADC using Open-Loop Amplifiers and Digital Calibration in 90nm CMOS," in *VLSI Circuits Symposium, Digest of Technical Papers*, June 2008, pp. 18–19.

79. C.-C. Huang, C.-Y. Wang, and J.-T. Wu, "A CMOS 6-Bit 16-GS/s Time-Interleaved ADC with Digital Background Calibration," in *VLSI Circuits Symposium, Digest of Technical Papers*, June 2010, pp. 159–160.

80. B. Murmann, "ADC Performance Survey 1997-2010," [Online]. Available: http://www.stanford.edu/~murmann/adcsurvey.html.

Index

A
Acquisition bandwidth, 66
Algorithmic behavior, 45
Analog-to-digital converter (ADC), 3
Architecture, 65, 70
Autocorrelation, 15, 17, 19, 22, 24, 29, 39, 41, 49, 50, 95

B
Background calibration, 37, 40, 45
Backplane, 1
Bandwidth limitations, 88
Bootstrapped track-and-hold, 66

C
Calibration ADC, 39, 40, 42–44, 49, 74
Calibration algorithm, 4, 38, 41, 45, 49, 74, 85
Calibration clock, 43, 44, 85
Calibration cycle, 48, 85
Calibration DAC, 71
Calibration engine, 72
Capacitive load, 76
Charge injection, 68
Clock distribution, 32, 60, 109
Clock-feedthrough, 68
Clock-gating, 44
CML latch, 54, 69
Common-mode, 55, 66, 77, 100
Comparator, 43, 54, 56, 62, 69, 71, 73, 99, 109
Comparator energy, 54, 56
Concavity constraints, 17
Convergence, 48, 85
Convergence speed, 45, 49
Conversion rate, 3

Cross-coupled inverter, 56, 69, 70, 73, 99
Crosscorrelation, 39, 40, 43

D
Data capture cards, 81, 83
Data rate, 1
Decimated output, 84, 85, 113
Delay, 32, 48, 75
Delay cell, 75
Delay line, 40, 48, 74, 76, 86
Deterministic bound, 17, 23, 27
Digital backend, 37, 39, 42, 43, 48
Digital calibration, 39
Digital domain, 3, 36
Digital processor, 36
Digitally-equalized serial link, 4
DNL, 84
DTFT, 7
Dynamic comparator, 54, 58, 69, 99
Dynamic performance, 84, 88
Dynamic throttling, 49

E
Encoder, 73
Energy efficiency, 4, 54, 69
ENOB, 90
Environmental changes, 37
Equalization, 1
Ergodic, 42, 51
Error analysis, 14
Error-rate, 73
Estimated jitter, 113, 114
Extrinsic capacitance, 57

M. El-Chammas and B. Murmann, *Background Calibration of Time-Interleaved Data Converters*, Analog Circuits and Signal Processing, DOI 10.1007/978-1-4614-1511-4, © Springer Science+Business Media, LLC 2012

F
Flash ADC, 4, 53, 65, 84, 109
FO4, 33, 75
FOM, 89
Foreground calibration, 37, 72
Fractional delay filter, 36
Frequency domain, 7, 8
Frequency response, 1
Full-adder, 73
Future work, 94

G
Gain, 10, 13, 18
Gate capacitance, 68, 76

H
Harmonics, 15, 88, 114

I
Impulse response, 25
INL, 84
Input capacitance, 53, 68, 70
Input sensitivity, 69
Input-referred noise, 70
Input-referred offset, 70–72
Inter-symbol interference, 1
Interconnect, 33
Interleaving factor, 12, 18, 31, 53, 57, 59, 62,
        107, 113
Intrinsic capacitance, 57
Inverse transform, 7
Iterative maximizer, 48, 85

J
Jitter, 22, 75, 88, 94, 114

K
Kickback noise, 70, 73
kT/C noise, 67

L
Latch gain, 69
Level converter, 77, 78
Linear time-invariant, 10, 23, 27
Linearized inverter, 56, 57
Linearized transistor, 99
LMS, 49
Load capacitance, 34, 57, 69, 75

Low pass filter, 23–25, 96
LSB, 72, 84
LVDS, 77
LVDS driver, 78

M
Matlab, 83, 84
Mean-square error, 15, 16, 18, 19, 22, 95
Measurement data, 81
Measurement results, 84
Metastability, 56, 59, 69
Mismatch, 4, 10, 17, 32
Mismatch limited, 17
Mixed-signal domain, 37
Monotonicity, 71, 77
Monte Carlo simulations, 23, 24, 33

N
NMOS switch, 66
Noise limited, 17
Nyquist, 88, 89

O
OCEAN, 105
Offset, 10, 12, 18, 41
Offset calibration, 72, 84
Ones-adder, 73, 109
Optimization, 53, 59, 62, 70, 99, 105
Output histogram, 84
Output spectrum, 9, 84, 88, 113

P
Performance limits, 58
Performance summary, 89
Phase generator, 32, 34, 77
Phase-shift, 8
PLL, 45, 77
PNOISE analysis, 70
Positive feedback, 69
Power consumption, 3, 37, 66, 70, 73, 75, 90
Power dissipation, 53, 57–59
Power divider, 82
Printed circuit board, 81
Proof-of-concept, 53
Propagation delay, 32, 33
Prototype ADC, 44, 65, 73, 78, 81, 92, 113

Q
QFN package, 81
Quantization, 17, 39, 41, 45, 50, 67, 111, 114

R

Regeneration, 54, 58, 69, 102
Residual error, 46
Residual timing skew, 88, 113
Resistor ladder, 59, 73

S

Sample-invariant, 49
Sampling capacitor, 66, 68
Sampling error, 19, 65, 88
Sampling switch, 34, 68
SDR, 66
Segmented DAC, 71, 76
Sense-amplifier, 69
Serial link, 1, 22, 53
Shift register, 77
Short-circuit current, 100, 102
Signal constraints, 38, 45, 49
Signal generator, 44, 81
Single sampler, 35
Skew correction code, 48, 49, 83
Spectral density, 22, 50
Static performance, 84
Stationarity, 49
Statistical bounds, 14, 24
Stochastic maximizer, 48, 49, 85
Sub-ADC, 5, 7, 10, 15, 22, 35, 39, 41, 43, 59,
  60, 65, 72, 77, 84
Subsampling, 8
Supply jitter, 76
Supply noise, 73, 75
Supply rejection, 75
Supply sensitivity, 73

T

Taylor series, 20, 109
Test setup, 81, 84
Thermal noise, 67, 72, 75, 114
Thin-oxide device, 68

Threshold voltage, 32, 99
Time constant, 55, 69
Time-interleaved ADC, 4, 5, 15, 17, 22, 23, 38,
  53, 59, 60, 62, 77, 88, 90, 99
Time-varying errors, 5, 10, 14, 15, 31, 36
Timing diagram, 44, 72, 109
Timing reference, 43
Timing skew, 10, 14, 19, 31, 35, 38, 40, 88, 113
Timing skew calibration, 38, 83, 84, 88, 92
Track-and-hold, 35, 65, 68, 112
Transconductance, 57, 69, 99
Trim DAC, 71
TSMC, 32, 81, 89

U

Uniform distribution, 56
Uniform sampling, 32

V

Variable capacitor, 75, 76
Variations, 32–34, 70, 71, 109
Vector space, 15
Voltage regulator, 75

W

Wallace Encoder, 73, 109
Wide-sense cyclostationary, 21, 29, 95
Wide-sense stationary, 15, 23, 49, 95
Wireline systems, 3

Y

Yield, 71, 75

Z

Zero-crossing, 51

Printed in the United States
by Baker & Taylor Publisher Services